MOON LANDING HOAX

MOON LANDING HOAX

QUINN SILVER

CONTENTS

1. Chapter 1: Introduction to the Moon Landing Contro — 1
2. Chapter 2: The Race to the Moon – Context and Moti — 9
3. Chapter 3: Key Figures in the Moon Landing Conspir — 17
4. Chapter 4: Analyzing the Apollo Footage and Photos — 25
5. Chapter 5: Space Suits, Equipment, and the Environ — 33
6. Chapter 6: The Van Allen Radiation Belts – A Techn — 41
7. Chapter 7: The Apollo Program – Financial and Logi — 49
8. Chapter 8: "Evidence" of Staging and Hollywood Con — 59
9. Chapter 9: Why Haven't We Returned? A Lingering Qu — 67
10. Chapter 10: Witnesses and Whistleblowers – Missing — 75
11. Chapter 11: Rebuttals from NASA and Scientific Com — 83
12. Chapter 12: The Psychological Appeal of Conspiracy — 91
13. Chapter 13: The Influence of Media and Pop Culture — 99
14. Chapter 14: Surveying Public Opinion – Belief in t — 107
15. Chapter 15: The Scientific Method vs. Conspiracy T — 115

16 Chapter 16: Final Thoughts – Why the Moon Landing 123

Copyright © 2024 by Quinn Silver
All rights reserved. No part of this book may be reproduced in any manner whatsoever without written permission except in the case of brief quotations embodied in critical articles and reviews.
First Printing, 2024

CHAPTER 1

Chapter 1: Introduction to the Moon Landing Contro

The Apollo Moon Landings – A Historical Milestone and Controversy

The Apollo moon landings in 1969 marked a defining moment not only for the United States but for humanity as a whole. When Neil Armstrong took his historic first steps onto the moon's surface on July 20, his words, "That's one small step for man, one giant leap for mankind," resonated around the world. This extraordinary event was broadcast on television screens globally, captivating millions and symbolizing the pinnacle of human ingenuity and determination. For many, the Apollo missions represented a triumph over the vast unknown, a story of bravery, science, and technology coming together to conquer a nearly unimaginable goal.

Amid the Cold War era's intense rivalry between the United States and the Soviet Union, the Apollo program offered the U.S. a chance to demonstrate its scientific prowess and technological superiority. It was the height of the space race, with each nation striving to be the first to accomplish milestones beyond Earth's atmosphere.

The Soviet Union's launch of Sputnik in 1957 had sparked both awe and fear in the U.S., igniting a fierce determination to outpace Soviet advancements in space. When the Apollo 11 mission succeeded in landing humans on the moon, it was a clear statement: America had not only caught up with but surpassed its Soviet competitors. The achievement reinforced a sense of national pride and was celebrated by people around the world who saw it as a testament to human potential.

However, not everyone accepted the moon landings as genuine. Almost immediately after the Apollo missions, a small yet vocal group of skeptics began to question the authenticity of NASA's achievements. Despite the wealth of footage, photographs, and testimonials from the astronauts, these skeptics pointed to what they saw as inconsistencies and irregularities in the evidence. While the vast majority of people celebrated the success, conspiracy theories about the moon landings slowly began to circulate, picking up momentum in the years that followed. Some argued that the technology needed to accomplish a moon landing simply didn't exist in 1969, while others pointed to what they believed were strange lighting and shadow effects in NASA's photos, suggesting studio lighting rather than sunlight in space.

The moon landing hoax theory has become one of the most enduring conspiracy theories of the modern era, captivating skeptics and curious minds alike. While the idea may seem far-fetched to many, a surprising number of people across generations continue to believe, or at least entertain the notion, that NASA may have staged the entire event. Skepticism toward the moon landings taps into a broader mistrust of government and authority, particularly when high-stakes events like the Cold War are involved. For some, the idea that the moon landings might be a carefully constructed deception is not only plausible but even probable. And with the rise of the internet, these ideas have only gained more visibility, enabling theorists and believers to share and amplify their views like never before.

Thus, the Apollo moon landings sit at a curious intersection of history and controversy, universally recognized yet questioned by a segment of society. This book seeks to explore this enduring conspiracy theory, examining why the Apollo missions still hold such fascination, how the hoax theory evolved, and what this belief says about our relationship with science, government, and truth itself.

Origins of the Moon Landing Hoax Theory

The idea that the Apollo moon landings were staged did not emerge overnight but took root slowly in the years following the missions. While the vast majority of people accepted the landings as real, a small but vocal community began to question the narrative. One of the most influential voices among the early skeptics was Bill Kaysing, whose 1976 book, *We Never Went to the Moon: America's Thirty Billion Dollar Swindle*, claimed that NASA had faked the Apollo moon landings. Kaysing's work is widely credited with popularizing the moon landing hoax theory and establishing a blueprint for many of the arguments still echoed by conspiracy theorists today.

Kaysing was a former technical writer for Rocketdyne, a company that manufactured engines for the Apollo program, and his background in the aerospace industry gave him a certain authority, even though he was not a scientist or engineer. In *We Never Went to the Moon*, Kaysing alleged that NASA lacked the technical capability to safely send astronauts to the moon and back. He argued that the agency, under intense pressure to beat the Soviet Union, decided to stage the landings rather than risk a highly public failure. To support his claim, Kaysing pointed to what he saw as flaws and irregularities in the Apollo footage and photos, including unusual lighting, unexplained shadows, and the famous image of the American flag appearing to wave in a supposedly airless environment.

Kaysing's book presented these "anomalies" as evidence that NASA had created a sophisticated ruse to deceive the public. He suggested that the footage was produced on a soundstage, possibly in a Hollywood studio or even in a hidden government facility in the Nevada desert. His arguments resonated with a segment of the public, particularly those who were already skeptical of government agencies or who felt alienated from mainstream narratives. Although *We Never Went to the Moon* was dismissed by the scientific community, it attracted attention and found an audience among those who saw it as a bold challenge to official history.

As Kaysing's ideas spread, others joined the chorus of skepticism, contributing their own theories and interpretations. Over time, the hoax theory gained momentum, particularly as other writers, filmmakers, and skeptics added layers to the narrative. These skeptics often cited Kaysing's work as foundational, using his arguments as a basis to claim that not only Apollo 11, but all six manned moon landings, were faked. This growing community began to find evidence in everything from technical documents to NASA's own images and video, constructing an alternative story of the Apollo program.

The 1980s and 1990s saw the moon landing hoax theory take on a life of its own, with some proponents going so far as to suggest that Hollywood director Stanley Kubrick was involved in creating the moon footage. Kubrick's 1968 film *2001: A Space Odyssey* had used groundbreaking special effects to depict space travel, and skeptics argued that his expertise in visual effects would have made him an ideal candidate to create convincing "moon landing" footage. Though entirely speculative, this narrative added intrigue to the theory, suggesting that NASA may have enlisted Hollywood's help to craft its supposed deception. In time, the theory became part of the broader landscape of government conspiracies, standing alongside narratives about cover-ups, secret agencies, and hidden agendas.

This evolving web of doubt, suspicion, and speculation has given the moon landing hoax theory an unusual durability. Decades after Kaysing's book was published, the theory continues to find new believers, fueled by a combination of historical mistrust, persuasive anecdotes, and a lack of scientific literacy in certain areas. Today, Kaysing's influence remains visible in online forums, documentaries, and books that continue to question one of humanity's greatest achievements. The origins of this theory reveal more than just a story of skepticism; they reflect a persistent undercurrent of doubt in society, one that challenges official narratives and finds fertile ground in those who feel excluded or disillusioned with established institutions.

Purpose, Scope, and Audience of This Book

Moon Landing Hoax is a comprehensive exploration of one of the most enduring conspiracy theories in modern history: the claim that the Apollo moon landings were faked by NASA and that humans never actually set foot on the moon. This book is designed to provide readers with an objective examination of both sides of the debate, shedding light on why this theory has persisted for over half a century and what it reveals about human curiosity, skepticism, and our collective relationship with authority and truth. Whether readers are seasoned skeptics, firm believers, or simply curious observers, this book seeks to engage with the nuances of the theory and the evidence, providing a balanced view of an extraordinary claim.

The purpose of *Moon Landing Hoax* is not to convince readers one way or the other but rather to unpack the elements that fuel the debate. From alleged inconsistencies in NASA's photos and videos to the technical limitations of the 1960s, each chapter examines key arguments put forth by hoax proponents and evaluates the scientific and historical responses from NASA, experts, and researchers. The intent is to respect both the appeal of skepticism and the foundation of scientific inquiry, encouraging readers to weigh the evidence and reach their own conclusions.

To achieve this balanced approach, the book is divided into carefully structured chapters, each focusing on a specific aspect of the moon landing hoax theory. Some chapters dive into the technical and scientific details, analyzing the supposed anomalies in the photos, videos, and equipment used in the Apollo missions. Others look at the psychological and sociological factors that contribute to belief in conspiracy theories, particularly the appeal of questioning accepted narratives. In later chapters, the book explores how popular culture and media have amplified the theory, ensuring its persistence in public consciousness even decades after the last Apollo mission. By separating the book into focused sections, readers can explore the theory from multiple angles, gaining a well-rounded understanding of why the idea of a faked moon landing remains so compelling to this day.

The intended audience for *Moon Landing Hoax* is broad, including anyone with an interest in space, history, conspiracy theories, or psychology. This book is designed to be accessible to readers of varying levels of expertise, from those with deep knowledge of space exploration to those simply interested in exploring alternative viewpoints. While the content is factual and based on credible sources, it is presented in a manner that invites open-minded inquiry, aiming to engage readers who may not be convinced of either side. It speaks to those who may be skeptical of "official" narratives, as well as those who respect scientific achievement but are intrigued by the persistence of conspiracy theories.

Ultimately, *Moon Landing Hoax* aims to do more than just revisit familiar arguments; it seeks to explore what drives people to believe in a story as improbable as faking a moon landing. Why does this particular theory, above others, hold such fascination for so many people? How does it challenge our assumptions about progress, trust, and truth? By engaging with these questions, the book endeavors to provide readers with a richer understanding of one of history's most remarkable achievements and the skeptical response it continues to provoke. Whether readers walk away with more questions or a deeper conviction in their beliefs, this book encourages a thoughtful and informed engagement with one of the most debated events of the 20th century.

CHAPTER 2

Chapter 2: The Race to the Moon – Context and Moti

The Cold War and the Birth of the Space Race

The Apollo moon landing can only be fully understood within the tense, high-stakes context of the Cold War, a period marked by political, ideological, and technological competition between the United States and the Soviet Union. Emerging from World War II as the world's superpowers, the U.S. and the Soviet Union engaged in an extended struggle for global influence and dominance, which played out on multiple fronts. The space race became one of the most visible arenas of this struggle, symbolizing not only technological prowess but also ideological supremacy. The launch of the Soviet satellite *Sputnik* in 1957 had a profound impact on American society and government, igniting fears that the U.S. was falling behind in technological advancements with potentially dire security implications.

The launch of *Sputnik* marked the world's first artificial satellite and shattered the illusion that the United States was uncontested in technological supremacy. To many Americans, *Sputnik* signaled the unsettling reality that the Soviet Union could now potentially launch weapons from space, creating a new and urgent dimension to the arms race. American officials quickly understood that the nation's position as a global leader depended on responding to this Soviet victory with its own technological leap. The U.S. was jolted into action, determined to regain dominance by advancing its own space program. Public perception of American strength was at stake, and the nation's reputation as a global leader hinged on whether it could meet or exceed the Soviet achievement.

To combat this perceived vulnerability, the U.S. government took swift and decisive action. In 1958, Congress passed the National Aeronautics and Space Act, establishing NASA as the agency responsible for coordinating and leading American efforts in space exploration. From its inception, NASA was given the monumental task of restoring U.S. prestige in the eyes of the world. This mandate was not only a scientific and technological endeavor but also a public relations campaign designed to reassure Americans that the country was still capable of innovation and achievement at the highest levels. The Soviet Union's early victories in space, such as the successful launch of *Sputnik* and later the flight of cosmonaut Yuri Gagarin, who became the first human to orbit Earth in 1961, placed further pressure on NASA and the U.S. government to respond with something even more extraordinary: landing a man on the moon.

The Cold War climate meant that the stakes for space exploration were extraordinarily high. For the U.S., losing the space race to the Soviets would signify more than a technological shortfall; it would be seen as a defeat of the American way of life. This was a battle of ideologies, with the democratic principles of the U.S. pitted against the communist doctrines of the Soviet Union. In this atmosphere, space exploration became both a symbol of national pride and a tool for propaganda. Americans rallied around the idea of exploring the unknown, inspired by the possibility of achieving what no other country had managed to accomplish. The race to the moon was no longer just about science; it was about proving the superiority of American ideals on a global stage.

This intense rivalry, driven by political and ideological goals, laid the groundwork for the Apollo missions. It also helps explain why some people later came to believe that the pressure to succeed at any cost may have tempted NASA to fabricate a moon landing. The stakes were higher than most Americans knew, with national prestige, technological reputation, and political power all bound up in the quest to reach the moon. Whether or not this pressure could have led to a staged landing is a question that remains at the heart of the moon landing hoax theory.

NASA's Role and the Urgency of a Moon Landing

In response to the Soviet Union's early victories in space, the United States moved quickly to consolidate its own space program, establishing NASA in 1958 with the mission of leading America's quest for supremacy beyond Earth. NASA became the focal point for the nation's ambitions in space, tasked with a seemingly impossible objective: to achieve a manned moon landing before the end of the 1960s. At the height of the Cold War, this mission was as much about national pride and geopolitical advantage as it was about scientific exploration. Under enormous pressure to demonstrate American superiority, NASA became not only a scientific institution but also a powerful symbol of America's will, ingenuity, and competitive edge.

In 1961, just months after Soviet cosmonaut Yuri Gagarin became the first human to orbit the Earth, President John F. Kennedy boldly set forth the United States' goal to land a man on the moon by the close of the decade. In his iconic address to Congress, Kennedy framed the mission as a test of American resolve, determination, and technical prowess. His speech galvanized the American public and established the moon landing as a crucial milestone for the country's identity. Kennedy's words resonated deeply with the American people, but they also set an ambitious and daunting deadline that placed extraordinary demands on NASA. The agency's leaders and engineers were now on a national stage, tasked with delivering a feat of science and technology under tight deadlines and intense scrutiny.

The Apollo program rapidly became one of the most costly and extensive projects ever undertaken by the U.S. government. An unprecedented amount of resources was allocated to ensure the program's success, with tens of billions of dollars invested and the best scientific minds recruited. NASA had to overcome a series of formidable engineering, logistical, and safety challenges, all while meeting the high expectations set by the government and the public. For the scientists, engineers, and astronauts involved, failure was not an option. Beyond the reputational costs, the failure of Apollo would have had serious political implications, potentially undercutting the U.S. in its ongoing struggle with the Soviet Union and damaging American morale. The project became not just a scientific mission but a matter of national security and prestige.

With the immense pressure to achieve a successful moon landing, the stakes grew higher, and the fear of failure became palpable. The U.S. government and NASA had committed not only to reaching the moon but also to doing so on a tight timeline, and there were no guarantees that success was within reach. The Apollo program's immense budget and high-profile setbacks, such as the tragic Apollo 1 fire in 1967, further underscored the challenges of the mission. These obstacles heightened the urgency, creating a climate in which the goal sometimes seemed nearly unattainable. Some hoax theorists argue that this intense pressure, coupled with the high political stakes, might have created an environment in which a staged moon landing seemed like a viable last resort to avoid national embarrassment.

From the perspective of hoax proponents, NASA's enormous undertaking—bolstered by public expectations and Kennedy's audacious vision—created a "make or break" situation. To those who subscribe to the hoax theory, the moon landing was not just about scientific discovery but about maintaining U.S. dominance in a heated ideological war. They argue that, faced with an unyielding deadline and no guarantee of success, the U.S. government might have considered fabricating the moon landing footage to protect its position on the world stage. By understanding the level of urgency, the massive investment, and the overwhelming expectations placed upon NASA, it becomes clearer why some believe the agency may have had a motive to stage what has been remembered as one of humanity's most celebrated achievements.

Pressure to Succeed – Theories on Faking the Landing

As the 1960s unfolded and the United States pushed its ambitious Apollo program forward, the stakes for a successful moon landing grew ever higher. The pressure on NASA was palpable, with each failure or setback amplifying the urgency of achieving a victory that could solidify America's technological dominance. To many in government, academia, and the public, the space race was no longer just a contest of rockets and astronauts—it was a litmus test for the credibility of the American way of life against Soviet communism. Failure to meet the challenge would not only tarnish America's image but could embolden Soviet confidence and weaken the U.S.'s standing as a global leader. This fierce backdrop set the stage for the development of theories that NASA, desperate to uphold its mandate, may have chosen an extraordinary path to guarantee success: staging the moon landing itself.

The theory that NASA might have faked the moon landing hinges on the premise that the U.S. government simply could not afford to fail. By the late 1960s, the political and public appetite for expensive space missions was beginning to wane. Domestic issues such as civil rights unrest and the escalating Vietnam War had captured the attention of the American public and policymakers alike, casting shadows on the ever-growing expenditures required to fulfill Kennedy's 1961 promise. According to proponents of the hoax theory, these conditions contributed to an atmosphere in which NASA would have faced tremendous pressure to deliver results, regardless of the obstacles.

The Apollo program had endured numerous challenges, including catastrophic events like the Apollo 1 tragedy, where three astronauts lost their lives in a command module fire during a pre-launch test. This incident raised doubts about whether NASA could overcome the multitude of technical problems required to ensure a successful moon landing. The failure, though tragic, served as a somber reminder of the human and political costs of the mission. If NASA were to fail publicly at a later stage—during a moon mission itself—the repercussions could extend far beyond technical setbacks; it could spell a damaging blow to national pride and political legitimacy. For theorists, this intense climate of risk and consequence provided fertile ground for speculation that, rather than accept the potential humiliation of failure, NASA would resort to an elaborate ruse to maintain its standing.

According to this perspective, a staged moon landing was seen as a strategic decision, a safeguard against failure under unimaginable pressure. Hoax proponents often suggest that the U.S. government, along with NASA, had the motive and resources to pull off such a deception. They argue that a fake moon landing, produced with cutting-edge 1960s film and special effects technology, would be far less costly and less risky than a failed mission. The perceived sophistication of the U.S. government and its agencies lent plausibility to the idea that such an operation could be carried out in total secrecy.

Moreover, the context of Cold War-era propaganda and psychological warfare lends some weight to this line of thinking. During that time, governments on both sides of the Iron Curtain engaged in highly coordinated disinformation campaigns to sway public opinion and outmaneuver their opponents. The notion that NASA might fabricate a moon landing as part of this psychological struggle fits within a broader pattern of Cold War strategies aimed at bolstering a nation's image and undermining that of its rivals. By achieving a moon landing—real or staged—the U.S. would reinforce its narrative as the pinnacle of human achievement and freedom, contrasting sharply with the Soviet image of oppressive control and secrecy.

These pressures and political contexts help explain why some people find the moon landing hoax theory plausible. It suggests that the moon landing wasn't just a milestone in space exploration but also a dramatic political maneuver where the stakes were higher than most Americans could comprehend at the time. While mainstream historians and scientists argue strongly in favor of the landings' authenticity, hoax proponents see a different story: one of desperation, spectacle, and calculated deception to win the ideological battle of the century. Whether seen as a testament to human ingenuity or a complex fabrication, the moon landing remains, in this theory, the ultimate Cold War triumph—one potentially written not in lunar dust, but on a Hollywood soundstage.

CHAPTER 3

Chapter 3: Key Figures in the Moon Landing Conspir

Bill Kaysing – The Catalyst for the Hoax Theory

Among the central figures who ignited the moon landing hoax movement, Bill Kaysing stands out as both a pioneer and an unlikely hero of skepticism. Kaysing, who had a background in technical writing and editing, worked for Rocketdyne, a company involved in producing engines for NASA's Apollo missions. Although his role did not involve direct engineering or scientific work, Kaysing's proximity to the space industry lent a veneer of credibility to his claims. His disillusionment with NASA and the space program led him down a path that would culminate in the publication of his controversial 1976 book, *We Never Went to the Moon: America's Thirty Billion Dollar Swindle*.

In his book, Kaysing laid out what he believed to be compelling reasons why the Apollo moon landings were an elaborate hoax. He argued that NASA lacked the technical capabilities to safely land a man on the moon and return him to Earth, citing perceived flaws in the spacecraft's design and the intense radiation of the Van Allen belts as insurmountable obstacles. According to Kaysing, the U.S. government, faced with the prospect of failure and the political consequences it would entail, orchestrated an elaborate deception to fulfill President Kennedy's ambitious promise of a manned moon landing by the end of the 1960s.

Kaysing's theory was not grounded in any insider knowledge of classified information but was based on his interpretations of publicly available data and personal conjecture. He pointed to supposed anomalies in NASA's moon landing photographs, such as the absence of stars in the lunar sky and what he claimed were inconsistencies in lighting and shadows, as evidence of tampering or fabrication. He argued that these images were staged in a studio, where lighting could be manipulated to create the illusion of a lunar landscape.

Despite the lack of verifiable proof, *We Never Went to the Moon* resonated with a segment of the public already prone to skepticism about government transparency, especially in the wake of scandals like Watergate. Kaysing's assertions found an audience willing to question the official narrative and consider that the U.S. government might have resorted to deceit to maintain national pride and global supremacy during the Cold War.

The publication of Kaysing's book was pivotal; it served as a catalyst for the development of a broader moon landing hoax movement that would persist for decades. His claims were taken up and expanded upon by other conspiracy theorists, each adding new dimensions to the arguments he first popularized. Kaysing's work laid the foundation for what would become a complex tapestry of doubt, bolstered by a mix of photographic analyses, scientific claims, and speculative reasoning.

While mainstream scientists and space historians dismissed Kaysing's theories as unfounded and riddled with inaccuracies, his role as the initiator of this widespread skepticism is undeniable. He planted the seed of doubt that would grow into a resilient conspiracy theory, one that has evolved and adapted with new voices and modern technology. Kaysing's book may not have convinced the scientific community, but it achieved a different kind of success: it sparked a debate that has endured long after his original publication, shaping a movement that continues to question the legitimacy of the Apollo moon landings to this day.

Other Influential Voices and Their Theories

While Bill Kaysing laid the groundwork for the moon landing hoax theory, he was far from the only voice to capture public attention. As his initial ideas began to circulate, other figures emerged, each contributing their own interpretations and pieces of "evidence" that added fuel to the growing skepticism. Two such figures who helped amplify and diversify the moon landing hoax narrative were Ralph René and Bart Sibrel.

Ralph René, an author and self-proclaimed independent researcher, became well-known for his book *NASA Mooned America!* published in the 1990s. René, who did not have a formal background in space sciences, took an approach steeped in meticulous yet unconventional scrutiny. His arguments often focused on physics and environmental conditions, emphasizing what he believed to be scientific impossibilities in the moon landing accounts. For example, René questioned how the astronauts could have survived passage through the Van Allen radiation belts without lethal exposure. He suggested that the technology of the 1960s was insufficient to shield human life from such extreme radiation, challenging NASA's claim that the astronauts made it through unscathed.

René also highlighted apparent discrepancies in photographs and videos released by NASA. He argued that certain shadows within the images did not align correctly, implying multiple light sources were used, which he claimed indicated artificial studio lighting rather than the singular, far-reaching light of the sun. René's interpretations of photographic "anomalies" such as inconsistent shadow directions and the absence of visible stars in the lunar sky were embraced by those already skeptical of official narratives. His book attracted readers who were inclined to see evidence of manipulation in the details, cementing his role as a prominent figure in the hoax community.

Another influential figure was Bart Sibrel, a filmmaker and outspoken critic of NASA's lunar missions. Unlike Kaysing and René, Sibrel took a more confrontational approach in his quest for answers. He produced several documentaries, most notably *A Funny Thing Happened on the Way to the Moon*, which aired his suspicions about the authenticity of the Apollo footage. In this film, Sibrel presented a compilation of supposed inconsistencies in the mission's documentation, including video segments he claimed showed astronauts staging scenes inside the spacecraft to mimic being far from Earth. Sibrel's approach was particularly incendiary; he sought to confront Apollo astronauts directly, famously engaging in heated encounters with figures like Buzz Aldrin, whom he accused of perpetuating a lie. These confrontations only served to elevate his profile and amplify the hoax theory, drawing both support and condemnation.

Sibrel's documentaries were characterized by dramatic presentations and persuasive narration, targeting the emotional as much as the rational. He claimed that NASA's reluctance to release certain raw footage or respond comprehensively to all allegations was indicative of a cover-up. While most scientists dismissed these accusations as misunderstandings or misinterpretations of technical details, Sibrel's work reached millions through VHS tapes and later online platforms, broadening the hoax theory's exposure to a new audience.

These influential figures shared common themes in their arguments: doubts about technological feasibility, inconsistencies in visual records, and perceived government motives for staging the landings. While Kaysing provided the foundational skepticism, René's scientific angles and Sibrel's provocative media tactics helped transform the moon landing hoax theory from fringe speculation into a sustained cultural phenomenon. Their voices, combined with the emerging accessibility of conspiracy material through print, broadcast, and eventually the internet, solidified the hoax movement as an enduring part of public discourse.

Despite overwhelming evidence supporting the authenticity of the Apollo missions, the combined efforts of Kaysing, René, Sibrel, and others created a narrative powerful enough to persist through generations. Each of these figures contributed to a larger tapestry of doubt, leveraging their interpretations of evidence and their unique communication styles to shape and spread a theory that still sparks debate and curiosity to this day.

Evidence and Arguments Presented by Key Figures

The various figures who championed the moon landing hoax theory, from Bill Kaysing to Ralph René and Bart Sibrel, constructed a foundation of arguments that have resonated with conspiracy theorists for decades. The "evidence" they cited, while often debunked by experts, was presented in ways that tapped into the public's fascination with uncovering hidden truths and questioning powerful institutions. This section examines the most prevalent pieces of evidence and the way these figures framed their arguments to suggest the moon landing was an elaborate ruse.

One of the most compelling aspects of the hoax theory lies in the purported photographic and video anomalies pointed out by figures like Kaysing and René. A common argument involves the absence of stars in the moon landing photos. Conspiracy theorists argue that if astronauts were on the moon, with no atmosphere to obscure the view, stars should be visible in the background. To hoax proponents, this absence indicated that the photographs were taken on a film set, where stars were either forgotten or deemed too difficult to accurately simulate. NASA, however, has consistently explained that the exposure settings on the cameras were adjusted for the brightness of the lunar surface, making the comparatively dim stars invisible. Nevertheless, the initial claim remains one of the most frequently cited pieces of evidence in the hoax narrative, embraced and disseminated by key figures.

Another major argument centers around inconsistencies in the shadows and lighting found in Apollo photographs. Hoax theorists such as René claimed that the angles of shadows in certain images were inconsistent with the singular light source of the sun, suggesting the use of multiple artificial lights. They pointed out photos where shadows of objects appeared to converge or diverge in a way that seemed illogical under natural sunlight. Experts in photography and physics have countered these claims by demonstrating how uneven lunar terrain can create such visual effects, with topography affecting shadow direction and length. Despite these explanations, proponents of the hoax continued to cite these anomalies as proof of studio staging, bolstering their arguments with selective images that appeared to support their case.

The technological capabilities of the 1960s were another focus for skeptics like Kaysing. He contended that the United States did not possess the means to safely land astronauts on the moon and return them to Earth. The Van Allen radiation belts became a central piece of evidence for this argument. Hoax theorists claimed that the radiation levels within these belts were so intense that any human passing through them would receive a fatal dose. NASA's response highlighted the fact that the Apollo spacecraft were designed with appropriate shielding and that the quick transit of the astronauts through the belts minimized their exposure to radiation. Nonetheless, the idea of lethal radiation exposure remained a popular argument, continually cited as proof that the moon missions could not have been real.

Bart Sibrel's contribution to the hoax theory included what he claimed were "smoking gun" pieces of evidence. One of his most famous assertions came from footage that he interpreted as showing astronauts staging shots of the Earth from space, using camera tricks to make it appear more distant than it actually was. In his documentaries, Sibrel narrated these clips with conviction, casting doubt on the integrity of NASA's footage. Mainstream experts explained the videos as misunderstood segments taken out of context, but Sibrel's dramatic storytelling captured the imagination of those willing to believe in a cover-up. This approach was effective at reinforcing the narrative of deceit that Kaysing, René, and others had built.

These arguments, while lacking in scientific validity, gained traction due to their simplicity and emotional appeal. They leveraged a blend of suspicion, selective evidence, and dramatic presentation to plant seeds of doubt. The legacy of these figures lies in their ability to transform technical details and photographic interpretations into a compelling story that suggested the possibility of grand deception. Even as scientific rebuttals have meticulously debunked these claims over the years, the evidence highlighted by Kaysing, René, and Sibrel continues to be part of the enduring fabric of the moon landing hoax theory, prompting discussions and skepticism long after the original Apollo missions.

CHAPTER 4

Chapter 4: Analyzing the Apollo Footage and Photos

Lighting and Shadows – The Core of the Hoax Argument

One of the most compelling arguments put forth by proponents of the moon landing hoax theory revolves around the alleged inconsistencies in the lighting and shadows within the Apollo mission photos and videos. To skeptics, these visual peculiarities serve as irrefutable evidence that the lunar scenes were staged using artificial studio lighting. By analyzing these claims, it becomes evident why the topic of lighting and shadows has become central to the debate and how experts in photography and physics have systematically countered these assertions.

A major point raised by hoax theorists is the claim that shadows in Apollo images often appear to diverge or intersect at odd angles, which they argue would be impossible if the only light source was the sun. For instance, in photographs of astronauts and lunar equipment, skeptics point to shadows that do not run parallel but instead seem to vary in direction. According to these theorists, this effect indicates the use of multiple light sources, akin to what might be found on a film set. If true, such an observation would undermine NASA's assertion that the photos were taken on the moon, where the sun should be the sole source of illumination.

NASA and photographic experts, however, have provided comprehensive explanations for these shadow anomalies. The moon's surface is uneven, marked by craters, rocks, and slopes that can distort the appearance of shadows. When shadows fall across such varied terrain, they may appear to be non-parallel or to bend, depending on the angle from which they are viewed or photographed. Additionally, the bright, reflective nature of the lunar regolith – the fine, powdery soil that covers the moon – plays a role. This reflective surface can scatter light and create what is known as secondary illumination, subtly brightening areas that might otherwise seem shadowed. These factors combine to create images that, while visually surprising, are consistent with what one would expect under lunar conditions.

Another key piece of supposed evidence is the apparent over-illumination of certain objects in photographs, such as astronauts or parts of the lunar module, even when they are positioned in shadow. Hoax proponents argue that if the moon's surface is illuminated only by sunlight, then areas in shadow should be significantly darker, and any object within them should be difficult to see. Critics suggest this discrepancy points to the use of artificial fill lights on a set. However, experts counter that the lunar surface's high albedo – its ability to reflect a significant amount of light – accounts for these effects. Sunlight bouncing off the lunar regolith can cast enough reflected light to brighten objects in shadow, explaining their visibility without the need for additional light sources.

These theories are further supported by extensive photographic and cinematic analyses conducted by professionals in the field. Experienced photographers note that similar shadow phenomena can be replicated on Earth when shooting in rugged landscapes with a single light source. The unique characteristics of lunar geography and the interaction of light with the moon's surface create conditions that are alien to most people's everyday experiences, making the resulting images appear suspect to the untrained eye.

Despite the scientific evidence debunking these claims, the shadow argument remains one of the most popular and enduring pieces of "evidence" cited by moon landing skeptics. The power of these visual anomalies lies in their immediate, instinctive appeal. To an observer unfamiliar with the moon's environment or the intricacies of light behavior, it is easy to interpret these images as artificially manipulated. This combination of visual complexity and public distrust fueled by decades of government secrecy and conspiracies makes such arguments difficult to dispel entirely, ensuring that the topic of lighting and shadows will remain a focal point in discussions of the moon landing hoax for years to come.

The "Waving Flag" Incident

Among the many pieces of evidence cited by moon landing hoax theorists, the so-called "waving flag" incident is perhaps the most iconic. The image of astronauts planting the American flag on the lunar surface, with the flag appearing to ripple as if blown by a breeze, is seared into the public consciousness. To skeptics, this moment is proof that the Apollo moon landings were filmed on Earth, where wind could cause such movement. To understand this argument's persistence and why it holds such sway in the popular imagination, it's essential to examine the footage, the scientific explanations, and the counterarguments presented by experts.

The heart of the waving flag argument lies in basic physics. In the vacuum of space, there is no air to create movement, so how could the flag flutter? This question has been seized upon by hoax proponents as evidence that the lunar landing was staged in an Earth-bound studio where an errant draft might cause a flag to move. Clips of the Apollo astronauts positioning the flag show it quivering and appearing to sway slightly, which conspiracy theorists interpret as clear evidence of a terrestrial setting. The argument hinges on the expectation that, absent wind or air, a flag on the moon should remain static.

However, NASA and scientific experts have provided a thorough explanation of the flag's behavior that aligns with the physics of the lunar environment. First, the flag used on the moon was specially designed to appear unfurled and stand upright in an environment with no atmosphere. A telescoping horizontal rod was inserted along the top of the flag to keep it extended. When astronauts planted the flagpole into the lunar soil, they twisted and adjusted it to ensure stability, inadvertently causing the fabric to oscillate. With no air to dampen the motion, the flag continued to move briefly after the astronauts let go, appearing to wave.

This continued motion is, in fact, more consistent with the properties of a vacuum than with an Earth-like atmosphere. On Earth, air resistance would cause the flag to settle quickly, but in the vacuum of space, there is no medium to slow down the movement. The flag's rippling was not due to wind but was the result of the momentum imparted by the astronauts during its placement. Experts in physics have demonstrated that such oscillations are precisely what one would expect in a low-gravity, airless environment like the moon's surface.

Another aspect that hoax proponents often overlook is that, when not being handled, the flag in the footage stands perfectly still, even when the astronauts are moving around it or the lunar module's engines fire up for departure. If a breeze were present on a studio set, the flag would have continued to flutter even after the astronauts stepped back. The flag's stillness in these moments underscores that its prior movement was not caused by wind but by the forces imparted by the astronauts themselves.

In addition to the scientific reasoning, video analyses and demonstrations by physicists and engineers have repeatedly shown that the behavior of the flag is consistent with what would occur in a vacuum. To illustrate this, experts have recreated similar experiments in vacuum chambers, where flags or objects suspended in airless environments behave similarly when set in motion.

Despite these clear explanations, the "waving flag" remains a powerful image and a potent symbol for those who doubt the authenticity of the moon landings. The visual simplicity of the claim – a flag appearing to move as if in a breeze – is compelling and easy for non-experts to grasp. This, combined with the broader context of governmental secrecy and public mistrust, helps explain why this incident has endured as one of the most cited points in moon hoax theories. Yet, as scientific evidence shows, the flag's behavior is perfectly aligned with the physics of the lunar surface, offering a powerful rebuttal to this persistent myth.

Other Photographic Anomalies and Their Rebuttals

Beyond the debated lighting discrepancies and the "waving flag" incident, moon landing hoax proponents have identified additional anomalies in Apollo mission photographs and footage as supposed evidence of a grand conspiracy. These include the absence of stars in the sky, reflections in astronaut helmets, and peculiarities in the visual details of the lunar surface. To understand the enduring nature of these arguments, it is important to explore both the claims made by hoax theorists and the scientific rebuttals that address each point.

One of the most frequently cited anomalies is the absence of stars in the photos taken by Apollo astronauts. Skeptics argue that, in the vacuum of space, with no atmosphere to obscure the view, stars should be vividly visible and punctuating the dark lunar sky. The lack of any starfield in the background is taken as evidence that the photos were staged in a controlled environment, where adding accurate constellations might have been deemed too difficult or inconsistent with known celestial maps. This argument capitalizes on the layperson's expectation that, in the darkness of space, stars must appear as prominent as they do in images taken on Earth at night.

However, the scientific explanation for this supposed anomaly is straightforward and well-supported. The cameras used by Apollo astronauts were set to capture images of the brightly lit lunar surface and astronauts in their reflective suits. The exposure settings on these cameras were adjusted to account for the harsh brightness of the moon's landscape, which is bathed in direct sunlight. Under these conditions, the relatively faint light from distant stars was not bright enough to register on film. This is comparable to taking a photograph on Earth during daylight; stars do not appear in such photos not because they are absent, but because the camera's exposure settings are optimized for the much brighter foreground. NASA's explanations, supported by professional photographers and scientists, align with what is known about photographic exposure in high-contrast environments.

Another point raised by hoax theorists concerns reflections visible in the visors of the astronauts' helmets. In some images, the reflection of the lunar module or the surrounding landscape appears in fine detail, leading conspiracy proponents to argue that certain objects, such as light sources or studio equipment, should not have been there. One famous claim suggests that the reflections include artificial lights or stage props used during filming. This argument feeds into the broader suspicion that the photos were staged in a studio environment.

Experts have countered this by analyzing the images and explaining the nature of reflections in curved, convex surfaces like helmet visors. Such reflections can capture a wide field of view and often distort what they reflect, making ordinary objects appear unusual or out of context. In the case of the Apollo images, the objects reflected were elements consistent with the lunar landing setup, such as the lunar module, other astronauts, and equipment. Far from being proof of studio trickery, these reflections are predictable and align with the panoramic view offered by the astronauts' helmets.

Lastly, skeptics often point to peculiarities in the shadows cast by objects on the lunar surface. Some argue that shadows of astronauts and equipment do not run parallel and appear to converge or diverge in ways that defy logic. This, they claim, suggests the use of multiple artificial light sources, which would indeed be needed on a film set but would be impossible on the moon. This argument has been addressed by physicists and visual experts who have pointed out that the uneven terrain of the moon can create illusions that disrupt the uniform appearance of shadows. The moon's surface is covered with craters, rocks, and slopes that can alter the path of a shadow, making it appear to bend or converge when viewed from certain angles. Additionally, the unique reflective properties of the lunar regolith can scatter light in unexpected ways, contributing to the visual complexity of the shadows.

In the end, these photographic and visual anomalies that skeptics present as proof of a hoax are, in reality, examples of how human perception and expectation can be at odds with the conditions of an alien environment. Scientific analyses and photographic expertise have consistently demonstrated that what appears to be suspicious or impossible is, in fact, in line with the known behavior of light, shadows, and reflection in the harsh, unearthly conditions of the moon. While these rebuttals are clear and well-documented, the allure of conspiracy theories persists, reminding us of the powerful role that mystery and skepticism play in shaping public dialogue.

CHAPTER 5

Chapter 5: Space Suits, Equipment, and the Environ

Technological Limitations of 1960s Space Suits

The idea that 1960s technology could not have produced space suits capable of surviving the harsh conditions on the moon is a cornerstone argument among those who believe the Apollo moon landings were a hoax. These skeptics point to the extreme temperature fluctuations, the exposure to cosmic and solar radiation, and the challenges of maintaining pressure and life support as factors that would have overwhelmed the technological capabilities of the era. To understand this argument and the counterpoints provided by experts, it's essential to delve into the history and engineering behind the Apollo space suits.

The space suits used during the Apollo missions, known as the A7L suits, were not mere garments but highly sophisticated pieces of technology. Constructed by the International Latex Corporation (ILC), these suits incorporated multiple layers of specialized materials to achieve durability, flexibility, and protection. The A7L suits consisted of up to 21 layers, each with a specific purpose: heat resistance, micrometeoroid protection, and maintaining internal pressure. The innermost layers provided comfort and cooling, using a water-circulation system that regulated body temperature, while outer layers were designed to reflect intense sunlight and insulate against extreme cold.

Hoax proponents argue that these suits were inadequate for withstanding the lunar environment, where daytime temperatures could soar above 250°F (121°C) and drop to -280°F (-173°C) at night. They claim that even with today's technological advancements, developing a suit that can endure such a range of conditions remains difficult, so achieving this in the 1960s would have been impossible. This argument, however, overlooks the specific timing and design parameters of the Apollo missions. NASA strategically planned the moonwalks during lunar mornings when the temperature was far less severe, thus avoiding the extreme highs and lows. The surface temperatures at these times, while still challenging, were within the range that the suits' layers of insulation could manage effectively.

Additionally, hoax proponents often cite concerns over radiation exposure, suggesting that astronauts would have been fatally exposed to cosmic and solar radiation without adequate protection. However, experts explain that short-term exposure during the brief duration of the moon missions posed a significantly lower risk than the continuous exposure astronauts would face during prolonged space travel or habitation. The A7L suits and the spacecraft itself were equipped with layers of material that provided sufficient protection for the limited time spent on the lunar surface. Furthermore, NASA carefully monitored solar activity to avoid periods of high solar radiation that could have endangered the crew.

The suits were rigorously tested in environments designed to mimic space conditions as closely as possible. These tests included vacuum chambers that simulated the pressure and temperature variances of space. Reports from astronauts who wore the suits during the missions affirm that they were effective; while the conditions on the moon were indeed harsh, the technology of the time was fully capable of providing the necessary protection. Engineers of the era had access to materials such as Mylar and Teflon, which were lightweight, strong, and had excellent thermal properties. These components were layered to optimize the suits' effectiveness without compromising the mobility astronauts needed to conduct their work on the moon.

Critics often underestimate the ingenuity and ambition that characterized the space race and NASA's Apollo program. The combination of cutting-edge materials, innovative design, and rigorous testing produced space suits that not only met but exceeded the requirements for lunar exploration. The A7L suits are now considered engineering marvels that set a benchmark for future space exploration. Far from being proof of a hoax, the successful performance of these suits stands as a testament to the technical prowess and determination of the 1960s space program.

The Lunar Module and Equipment Design

The Apollo lunar module, known as the Lunar Excursion Module (LEM), remains one of the most recognizable symbols of the moon landing era. This peculiar-looking, spindly craft was designed to accomplish one of the most daring feats in human history: landing on the moon's surface and returning to orbit. Despite its critical role in the Apollo missions, hoax proponents have frequently pointed to the LEM and associated equipment as evidence that the moon landings were staged. They argue that the module's seemingly fragile design and rudimentary technology indicate it could not have successfully functioned in the unforgiving environment of space and the lunar surface. However, a closer examination reveals that the LEM was, in fact, an engineering triumph that embodied the pinnacle of 1960s technological capabilities.

At first glance, the LEM's thin, metallic structure might appear ill-equipped to handle the harsh conditions of space and lunar gravity. The module's exterior was composed of layers of Mylar and Kapton foil, materials that appeared lightweight and flimsy compared to conventional aircraft or spacecraft. This design choice, however, was not due to a lack of resources or understanding; it was a deliberate and ingenious solution to a complex engineering problem. Weight was a critical factor for any component that needed to be launched, and every ounce carried into space required exponentially more fuel and power. Thus, the LEM was built to be as light as possible while still maintaining structural integrity.

The thin, gold-foil-like material seen on the LEM's surface served as an effective thermal insulation barrier, reflecting intense solar radiation and protecting the module from extreme temperatures. The craft was engineered specifically for the moon's environment, which lacks an atmosphere and subjects objects on its surface to direct, unfiltered sunlight and the deep cold of space. The LEM's dual-stage system—a descent stage for landing and an ascent stage for returning to the command module—was a novel approach that allowed NASA to create a craft optimized for efficiency, sacrificing robustness for practicality since the LEM only needed to operate on the moon's low gravity.

Hoax proponents often claim that the LEM's rudimentary-looking construction, including its visible rivets and patchwork-like surface, indicates that it could not have handled the stress of space travel or lunar operations. This argument fails to consider that the LEM was only meant to function in the low-gravity environment of the moon, which is one-sixth that of Earth's. The structural requirements for operation in such conditions are far different from those of Earth-based or orbital spacecraft. The design's minimalism was intentional, allowing it to perform its specific tasks: navigating the moon's surface, facilitating astronaut activity, and achieving takeoff from the moon.

One of the most impressive aspects of the LEM was its advanced computer system, which was cutting-edge for the time. The Apollo Guidance Computer (AGC) managed navigation and control tasks and worked in tandem with the astronauts. While it might seem archaic by today's standards, with computational power far less than that of a modern smartphone, it was revolutionary in the 1960s. The AGC's programming was sophisticated enough to handle real-time adjustments and problem-solving, as famously demonstrated during Apollo 11's descent, when Neil Armstrong took manual control due to computer warnings and guided the LEM to a safe landing.

To address these concerns, aerospace engineers and historians have highlighted how the LEM was extensively tested, both on Earth and through unmanned and manned missions, to ensure it could perform as intended. The Apollo 9 and 10 missions, in particular, served as critical tests of the LEM's capabilities before the historic Apollo 11 landing. These missions verified that the module's engines, structure, and life-support systems could operate as designed. The ascent stage, which took the astronauts back to the command module, used a reliable engine that had been tested repeatedly and incorporated redundancies to ensure a high likelihood of success.

The notion that the lunar module was inadequate for its purpose underestimates the genius of the engineers at NASA and its contractors. The LEM, far from being a smoking gun for conspiracy theorists, is celebrated as an innovative and practical solution to one of the most complex challenges in space exploration. The Apollo program's success was built on meticulous testing, incremental improvements, and an unwavering commitment to overcoming the engineering hurdles of the day.

The Moon's Environment and Technological Feasibility

The moon's environment is one of stark contrasts and unforgiving extremes, presenting significant challenges for any technology attempting to survive and function there. Without an atmosphere to moderate temperatures or shield against cosmic and solar radiation, the lunar surface swings from scorching heat to bitter cold, depending on the sun's position. The temperature can range from 250°F (121°C) in direct sunlight to -280°F (-173°C) during lunar night. Additionally, the surface is bombarded by micrometeoroids—tiny, high-speed particles that pose serious risks to equipment and astronauts alike. Hoax proponents argue that NASA's 1960s technology was inadequate to protect astronauts from these hazards, suggesting that the successful completion of the Apollo missions points to a staged event. However, a closer look at NASA's approach to these challenges reveals how the technology of the time, while pioneering, was meticulously engineered to address these very obstacles.

One of the most significant environmental challenges on the moon is its temperature extremes. The Apollo missions were timed to occur during the lunar day, shortly after sunrise at the landing site, to take advantage of favorable conditions. This planning helped ensure that temperatures, while still severe, were manageable with the thermal protection built into the space suits and the lunar module. The outer layers of the Apollo space suits were specifically designed with reflective and insulating materials that shielded astronauts from the intense heat of the sun. The multi-layered construction of the suits included aluminized Mylar and Dacron, which helped maintain a consistent internal temperature for the astronauts.

The lunar module itself incorporated a carefully considered thermal control system. Its exterior was covered with reflective insulation, such as Kapton and aluminized Mylar, which effectively minimized heat absorption. The thin, metallic foils seen in photographs might look inadequate to the layperson, but they were part of an advanced multi-layer insulation system that deflected solar radiation and stabilized the module's internal temperature. This lightweight approach was essential for ensuring that the lander could function as intended without exceeding the weight limitations imposed by the Saturn V rocket.

Another argument put forth by skeptics concerns radiation exposure. The moon lacks the magnetic field and atmosphere that protect Earth from cosmic rays and solar radiation, making astronauts theoretically vulnerable to harmful exposure. However, experts clarify that the brief time spent on the moon—averaging about 21 to 25 hours outside the lunar module for Apollo 11, for instance—resulted in minimal exposure. Radiation levels were closely monitored, and astronauts wore dosimeters to track their exposure. The doses they received were well below dangerous thresholds, thanks to the missions being planned to avoid periods of high solar activity. Furthermore, the lunar module provided adequate shielding during the time astronauts were inside it, using a combination of the craft's materials and strategic mission timing.

Micrometeoroids, another concern, posed a risk not only to astronauts but also to their equipment. These tiny space projectiles travel at extremely high speeds and can puncture surfaces on impact. The design of the lunar suits included several protective layers that could withstand impacts from smaller micrometeoroids, and the astronauts were trained to minimize the time spent exposed. The lunar module itself was built with a double hull and reinforced sections to handle the threat of micrometeoroid damage, contributing to a safer environment when the astronauts were inside.

Despite these comprehensive measures, hoax proponents argue that the scale of the technological challenges suggests that faking the moon landing would have been more feasible than overcoming them. However, such arguments overlook the sheer breadth of NASA's testing protocols, engineering prowess, and incremental technological advancements. From vacuum chamber tests that replicated space-like conditions to high-altitude simulations that pushed equipment to its limits, every component used in the Apollo missions was subjected to rigorous validation processes. These tests demonstrated that the technology, while simple by today's standards, was sufficient for short-term lunar missions.

The Apollo program's success was not only a triumph of ambition but also a testament to human ingenuity and the meticulous attention to detail that characterized NASA's approach. The technology of the 1960s was not without its constraints, yet it was specifically engineered to meet the demands of the lunar environment. The engineers, scientists, and astronauts involved in the Apollo missions worked within these constraints, applying groundbreaking knowledge to create a system that could survive the moon's unique challenges. This combination of strategic planning, innovative design, and rigorous testing provides a compelling counterpoint to the notion that the moon landings were fabricated. The technological feasibility of the Apollo missions was not an impossible feat; it was an extraordinary accomplishment that showcased humanity's capacity for solving even the most daunting problems.

CHAPTER 6

Chapter 6: The Van Allen Radiation Belts – A Techn

Understanding the Van Allen Radiation Belts

The Van Allen radiation belts are a formidable feature of Earth's magnetosphere, consisting of charged particles trapped by the planet's magnetic field. First discovered in 1958 by physicist James Van Allen during the Explorer 1 mission, these belts became a defining challenge for space exploration. Stretching thousands of miles above Earth's surface, the belts consist of two main layers: an inner belt, extending from about 400 to 6,000 miles above Earth, and an outer belt that ranges from approximately 8,000 to 37,000 miles above the planet. Both are filled with high-energy protons and electrons capable of inflicting serious damage on living tissue and electronics, making safe passage a complex issue for space missions.

The inner belt poses the greatest radiation threat due to its concentration of energetic protons. The outer belt, while broader and more variable in intensity, also presents significant exposure risks, particularly when solar activity enhances radiation levels. This understanding has led many conspiracy theorists to argue that crossing these belts in 1969, with the technology available at the time, would have been not just risky but lethal to astronauts. The concern is that prolonged exposure to this radiation would have posed severe health risks, potentially resulting in acute radiation sickness or long-term damage such as cancer.

For many who question the moon landings, the existence of these radiation belts serves as a linchpin for the hoax argument. The reasoning follows that if the radiation levels were truly as severe as described, the Apollo missions must have faked their lunar journeys because the technology to adequately protect astronauts from such hazards did not exist. This argument is persuasive to some, as the idea of space radiation conjures images of invisible, inescapable forces capable of undermining even the most meticulously planned missions.

To understand why these arguments persist, it is important to recognize the real challenges that the Van Allen belts present. The radiation within these belts is indeed hazardous; it can destroy unprotected circuits and electronics, degrade materials, and pose serious health risks to humans. Cosmic rays and solar particles can bombard objects within these belts at high speeds, creating conditions that require thorough risk assessment and mitigation strategies. This backdrop explains why the Van Allen belts became such a focal point for conspiracy theories that insist space travel in the late 1960s was either impossibly ambitious or outright deceptive.

The importance of these belts extends beyond Apollo; they remain a consideration for all manned and unmanned space travel. Satellites and other spacecraft must be designed with protection against radiation, balancing the need for effective shielding with weight constraints. The early space race, which unfolded in an era of fervent Cold War rivalry, pushed scientists and engineers to confront these challenges with unprecedented urgency and ingenuity.

Understanding the true nature of the Van Allen radiation belts helps frame the discussion about whether NASA's achievements were technologically feasible in the context of the 1960s. These belts are indeed a formidable part of Earth's space environment, and mitigating their effects required careful planning. However, as subsequent sections will detail, the approaches used to minimize exposure for Apollo astronauts were grounded in sound scientific principles and engineering solutions that had been developed over years of research and practical application.

Scientific Explanations for Safe Passage

The scientific and engineering communities of the 1960s were acutely aware of the challenges posed by the Van Allen radiation belts and developed careful strategies to ensure the safe transit of astronauts. One of the most significant factors that contributed to the success of the Apollo missions was the strategic planning of the spacecraft's route. The Apollo trajectory was deliberately chosen to pass through the thinnest parts of the radiation belts, minimizing exposure time and radiation dosage. This approach allowed the spacecraft to travel through the belts in a matter of minutes rather than hours, substantially reducing the total radiation absorbed by the crew.

Speed played a crucial role in mitigating radiation exposure. The Apollo spacecraft traveled at such a high velocity that the time spent within the most dangerous regions of the belts was kept brief—typically under an hour in total for both entry and exit. This swift passage was essential to minimizing radiation exposure to a level that would not endanger human health. The entire Apollo mission profile was designed to ensure that astronauts experienced only short-term exposure, which was calculated to be within safe limits.

Moreover, the protective measures incorporated into the design of the Apollo command module were tailored to provide adequate shielding against the radiation encountered. The command module's structure included layers of aluminum and other materials that offered significant protection against high-energy particles. Although the shielding was not sufficient for prolonged stays in a high-radiation environment, it was more than adequate for the relatively brief transits through the Van Allen belts. The exposure levels were closely monitored and assessed, and calculations confirmed that astronauts would receive a total dose well below the thresholds for radiation sickness or long-term harm.

Data and analysis from NASA, corroborated by independent space scientists, consistently support the conclusion that the total radiation dose received by the Apollo astronauts was comparable to the amount accumulated during routine medical X-rays. During the missions, astronauts wore dosimeters to measure their radiation exposure in real time. The results showed that, despite passing through the Van Allen belts, the recorded radiation exposure was far from the lethal levels suggested by some conspiracy theorists. These findings are reinforced by the health of the Apollo astronauts, most of whom lived for decades after their missions without experiencing the radiation-related illnesses predicted by skeptics.

The successful passage through the Van Allen belts was also facilitated by a deep understanding of space weather. NASA meticulously timed missions to avoid solar storms and periods of increased solar activity that could have heightened the radiation threat. Space agencies today continue to use similar strategies for manned and unmanned missions, employing real-time solar monitoring and predictive models to ensure safe space travel. This practice was already being developed in the 1960s, showing that NASA's approach to radiation hazards was based on sound, forward-thinking scientific methods.

Critics often overlook the extensive research and simulations that preceded the Apollo missions. The space agency conducted ground-based experiments using high-energy radiation environments to test spacecraft materials and shielding. The results of these tests informed the final design of the command module and proved that even the relatively basic shielding of the time could offer ample protection for short-term lunar missions. These preparations were complemented by unmanned test missions that provided crucial data about space radiation and validated the safety of the Apollo mission profile.

In summary, while the Van Allen radiation belts are indeed a significant feature of Earth's magnetosphere, they did not present an insurmountable barrier to the Apollo missions. The careful navigation through lower-radiation zones, the rapid transit through the belts, and the protective measures in the spacecraft all combined to ensure that the astronauts were safe. The exposure levels were well understood, planned for, and managed within acceptable limits. The success of these measures is evident in the fact that the Apollo missions were completed without incident related to radiation, providing a testament to the feasibility and scientific rigor of NASA's mission planning and execution.

Counterarguments and Hoax Proponents' Claims

Despite the wealth of scientific evidence supporting the Apollo missions' safe navigation through the Van Allen radiation belts, conspiracy theorists remain steadfast in their belief that NASA could not have accomplished such a feat with 1960s technology. Central to these claims is the argument that shielding technology at the time was insufficient to protect astronauts from lethal doses of radiation. Proponents of the hoax theory often cite exaggerated or misunderstood assessments of the radiation intensity within the belts to bolster their assertions that NASA's explanations were implausible.

A key argument presented by skeptics involves the perceived inadequacies of the materials used in the Apollo command module. Conspiracy theorists argue that thin aluminum and other light metals would not have been enough to prevent radiation from penetrating the spacecraft, leaving the astronauts vulnerable. They suggest that, given the purportedly high levels of radiation, exposure would have led to immediate radiation sickness or, at the very least, serious long-term health consequences. This assertion, however, misrepresents both the nature of the Van Allen belts and the levels of radiation involved during a rapid transit.

One common misconception among hoax proponents is that the Van Allen belts are uniformly deadly, without safe corridors for passage. In reality, as outlined by space scientists, the radiation levels vary significantly throughout the belts. The outer belt, for instance, is less dense than the inner belt, and both belts have regions with lower radiation concentrations. NASA exploited these variances to chart a course that passed through these less intense regions, allowing the Apollo spacecraft to avoid the highest concentrations of radiation. Additionally, the notion that prolonged exposure is necessary for severe radiation damage is often overlooked. The Apollo missions were designed to pass through the belts quickly, limiting exposure time to well below the levels that could cause acute harm.

Hoax advocates frequently cite data out of context, comparing the Apollo missions' brief exposures to prolonged stays in high-radiation environments, such as those encountered by satellite components left for years in space. However, the comparison fails to account for the relatively low total radiation dose absorbed during a short-duration flight through the belts. Radiation exposure is measured in units of sieverts or rems, with space experts clarifying that the doses experienced by Apollo astronauts were within non-lethal limits. The actual exposure, monitored with dosimeters, confirmed that the levels were equivalent to those faced by airline pilots on long-haul flights over the course of their careers—significant, but not deadly.

Critics also argue that radiation shielding would have been prohibitively heavy for a spacecraft to carry. While it is true that creating a fully radiation-proof capsule would be impractical due to weight constraints, NASA's engineers used a layered approach that balanced protection with the limitations of the Saturn V rocket's payload capacity. The command module's structure was reinforced with aluminum and other specialized materials capable of blocking particle radiation for the brief period necessary. Additionally, the spacecraft's orientation and the astronauts' position within it provided an extra layer of indirect shielding.

The health outcomes of the Apollo astronauts themselves offer compelling evidence against the hoax theory. Far from showing signs of radiation sickness, the astronauts reported no such symptoms and later medical studies found no unusually high rates of cancer or other radiation-linked conditions. These findings are consistent with the projected levels of radiation exposure from their transit through the Van Allen belts. Conspiracy proponents have struggled to counter these facts convincingly, often resorting to claims of orchestrated cover-ups or fabricated medical data.

Understanding the truth about the Apollo missions requires distinguishing between valid scientific concerns and arguments fueled by selective evidence and skepticism. The challenges posed by the Van Allen belts were real, but NASA's solutions—employing well-researched strategies for route selection, protective spacecraft design, and mission timing—were more than adequate. By addressing these challenges with a blend of engineering ingenuity and thorough risk assessment, the Apollo program demonstrated that even formidable obstacles like the Van Allen radiation belts were not insurmountable. This blend of preparation and science stands as a powerful counterpoint to claims that such missions could only have been staged.

CHAPTER 7

Chapter 7: The Apollo Program – Financial and Logi

The Massive Cost of the Apollo Program

The Apollo program was one of the most ambitious and expensive endeavors ever undertaken by the United States, a project that would redefine the limits of human achievement while becoming a symbol of Cold War competition. Between 1960 and 1973, the cost of the program reached an estimated $25.4 billion, a staggering figure for its time, equivalent to over $150 billion today when adjusted for inflation. The Apollo program represented approximately 2.5% of the total federal budget during its peak years, a level of spending that reflected both the high stakes of the space race and the U.S. government's determination to establish dominance over the Soviet Union.

The financial commitment to Apollo was motivated by more than just a desire to reach the moon—it was about showcasing technological and political superiority. The United States had fallen behind in the space race when the Soviet Union launched Sputnik in 1957 and subsequently sent Yuri Gagarin into orbit in 1961. The urgency to regain the upper hand was palpable, and President John F. Kennedy captured the public's imagination when he announced in 1961 that the nation would land a man on the moon and return him safely to Earth before the decade's end. This ambitious vision required an unprecedented allocation of resources, which Congress approved, motivated by the Cold War's fierce geopolitical competition.

However, the colossal expense did not come without scrutiny. Skeptics both then and now questioned the practicality and necessity of such an investment, especially when domestic issues like civil rights struggles and the Vietnam War demanded attention and resources. The financial strain of Apollo led to debates over whether NASA could sustain the support needed to continue its operations or whether it might have been tempted to justify its budget by other means.

These financial pressures are a cornerstone in conspiracy theories that argue NASA faked the moon landings. According to proponents of this view, the sheer cost of the project created a risk that failing to achieve its stated goals would result in public outcry, budget cuts, and a loss of national prestige. To avoid such consequences, theorists claim that NASA may have orchestrated a simulated moon landing to present the illusion of success. The idea of a massive financial cover-up suggests that the U.S. government was willing to go to great lengths to convince both its own citizens and the rest of the world that it had triumphed in the space race.

While these theories persist, they often overlook key details that support the program's legitimacy. Documentation, photographic evidence, testimonies from engineers, and the detailed logs of thousands of employees who worked on the project all point to a real and genuine effort. Yet, the staggering financial commitment remains a central argument for those who suggest that the incentive to stage the moon landings was strong.

Understanding the financial stakes of the Apollo program helps illuminate why some find it plausible that NASA might have resorted to deception. The pressure to succeed, combined with the national and international spotlight, created an environment where failure was seen as unacceptable. To hoax theorists, this context bolsters the narrative that the United States could not afford the risk of public or political backlash from a failed mission. However, for NASA, the investment represented the fulfillment of a bold promise—a commitment to demonstrating American ingenuity and leadership on a scale never before attempted.

Logistical Challenges of a Lunar Mission

The sheer scale of the Apollo program was unprecedented in both ambition and complexity. Sending astronauts to the moon and returning them safely to Earth was not just a test of human will; it was an extraordinary technical and logistical undertaking. This audacious goal required the development of technologies and processes that had never before been attempted. From the design of the spacecraft to the management of ground operations, every aspect of the Apollo missions posed formidable challenges. These logistical hurdles, which NASA managed to overcome, have become a focal point for conspiracy theorists who argue that the difficulties were so great that the missions must have been fabricated.

One of the most significant challenges was the development of the Saturn V rocket, the most powerful launch vehicle ever constructed. Standing 363 feet tall and generating 7.5 million pounds of thrust, Saturn V was a technological marvel but also a logistical nightmare. Its creation involved thousands of engineers and scientists working across the United States, coordinating efforts that pushed the boundaries of contemporary engineering. The rocket's design required innovations in propulsion, fuel management, and structural integrity to lift the payload necessary for a moon mission. The sheer number of tests, failures, and redesigns that marked the Saturn V's development underscored how ambitious the project truly was. For conspiracy theorists, these complexities feed the argument that the technological demands were too immense to overcome in the 1960s, suggesting that it was easier to fake the entire operation than to complete it successfully.

Aside from the rocket, the Lunar Module (LM) posed another set of formidable challenges. This spacecraft, responsible for landing astronauts on the lunar surface and then launching them back into orbit to rendezvous with the command module, required lightweight construction paired with reliable propulsion and life-support systems. The LM's design was so groundbreaking that even small errors could spell catastrophe. Every step, from assembling the module to integrating its systems, was fraught with potential failure points. Critics of the moon landing often point to this complexity as evidence that a real mission would have been too risky to attempt, considering the limited computing power and materials available at the time.

Astronaut training was equally intricate, involving simulated moon walks, zero-gravity practice sessions, and emergency drills that would prepare the crew for every conceivable challenge they might face. The rigorous nature of this training was necessary, given the unknowns that came with stepping onto the moon's surface for the first time. Astronauts had to be equipped to handle everything from equipment malfunctions to unexpected physical challenges, including the low-gravity environment and high temperature fluctuations. To conspiracy theorists, this exhaustive training regimen might seem like an elaborate charade designed to support a simulated event rather than a genuine preparation for a lunar landing.

Navigation and communication were also critical obstacles. The Apollo Guidance Computer (AGC), with its groundbreaking software and hardware, had to ensure precise trajectory calculations and control functions for both the command module and the LM. With just 64 kilobytes of memory, the AGC was limited by today's standards but represented an astonishing achievement in its time. The success of navigating hundreds of thousands of miles through space required not only the AGC but also a network of ground-based support that monitored the spacecraft's position and condition. For hoax proponents, the reliance on such "primitive" technology is cited as a reason to doubt the authenticity of the missions. They argue that NASA lacked the computational power to accomplish such complex navigation and control tasks, pointing to potential gaps in the technology narrative.

The logistical demands of the Apollo program extended beyond the engineering and training of astronauts to the management of resources and timing. With missions coordinated down to the minute, and every phase of the operation requiring precise execution, the challenge of synchronizing such an effort was immense. Critics argue that any small mistake could have derailed the mission, leading them to believe that NASA might have opted for an elaborate ruse instead of risking failure. This belief, however, disregards the countless hours of preparation and the lessons learned from earlier programs like Mercury and Gemini, which laid the groundwork for Apollo.

Despite these challenges, the Apollo program moved forward, achieving what many thought impossible. The successful moon landing was a culmination of hard-won expertise and innovation, marking a triumph of coordination and engineering. The logistical feats accomplished during this period remain unmatched in their scope and ambition, standing as a testament to the capabilities of human ingenuity under pressure. While the complexity of the Apollo missions continues to fuel doubts among skeptics, these very challenges, when examined in detail, underscore the dedication and meticulous work that made the moon landing a reality.

Theories Around Staging the Moon Landing for Funding and Prestige

As the Apollo program surged forward with relentless momentum, its soaring budget became a double-edged sword. On one hand, it underscored the United States' commitment to prevailing in the Cold War space race; on the other, it created immense pressure to succeed, no matter the cost. Conspiracy theorists have seized upon this financial strain to suggest that NASA, facing the daunting possibility of a failed mission, may have orchestrated a hoax to secure continued funding and preserve national prestige. According to this line of thinking, faking the moon landing would have been a calculated move to maintain the United States' image as the leader of space exploration and technological innovation.

The argument hinges on the belief that the Apollo missions were too critical to fail—politically, financially, and symbolically. By the late 1960s, the space race was more than a contest of scientific achievement; it had become a litmus test for the superiority of democracy versus communism. President Kennedy's famous 1961 declaration to send a man to the moon and return him safely to Earth before the end of the decade was not just a promise but a challenge that resonated deeply within the American psyche. The implications of failing to meet this challenge were profound: it would signify a colossal waste of taxpayer dollars and deal a blow to American prestige, which could have geopolitical ramifications.

Proponents of the moon landing hoax theory argue that the perceived risk of such a failure was too great, leading NASA and government officials to purportedly devise an alternative: staging the moon landing in a controlled environment. To these theorists, the continued influx of funds and public support were essential not only for Apollo but for NASA's broader mission. Without the moon landing as a crowning achievement, funding for future projects could have been jeopardized, potentially leading to the dismantling of NASA's ambitious space agenda.

In this context, conspiracy theorists often point to the circumstances surrounding NASA's funding and the growing skepticism of the American public as evidence of a motive. The space agency's budget peaked in 1966 and then began a gradual decline as competing national priorities, such as the Vietnam War and social programs, drew more attention and resources. Public interest in the space program, while initially fervent, started to wane as the cost ballooned and tangible results seemed elusive. A failed mission would have only exacerbated these issues, making it tempting, according to hoax proponents, for NASA to simulate success rather than risk a costly failure.

Yet, while the narrative of staging a moon landing to secure funding is compelling in its simplicity, it falters under scrutiny. Documented evidence points to a different reality—one of painstaking preparation, technological breakthroughs, and incremental success. Thousands of engineers, technicians, and scientists contributed to Apollo's intricate operations, leaving behind an extensive paper trail of research, testing, and problem-solving. Furthermore, Apollo 11's success was followed by subsequent missions that showcased advancements and learned lessons, such as Apollo 12's precise landing near the Surveyor 3 probe and Apollo 14's mastery over earlier navigation difficulties.

The rebuttal to funding-driven conspiracy theories also lies in examining the sheer scale of what would have been required to fake such a feat. Producing convincing footage, developing plausible scripts, managing thousands of workers involved in the Apollo program, and maintaining secrecy over decades are challenges of a magnitude that undermines the practicality of such a scheme. While theorists claim that only a small, elite group needed to be in on the hoax, this contention does not account for the hundreds of thousands of individuals who played essential roles across various NASA centers, contractors, and independent scientists around the world. To suggest that all of these people were either complicit or deceived stretches the bounds of plausibility.

Nevertheless, the theory that financial and political pressures could have driven a hoax persists, fueled by a cultural distrust of governmental institutions and a fascination with grand deception. The possibility of faking a moon landing offers an attractive narrative: it frames the U.S. government as cunning and capable of weaving an elaborate tale to maintain its image, a notion that taps into broader concerns about transparency and power. But when the Apollo missions are scrutinized alongside the overwhelming evidence of their authenticity, from lunar rock samples analyzed by geologists to independent radio amateurs tracking the spacecraft, the hoax theory becomes increasingly untenable.

In the end, while funding and national prestige were undeniable motivators behind the Apollo program, they do not support the notion of a hoax but rather reinforce the sheer determination that drove the United States to achieve what was once thought impossible. The legacy of Apollo is one of true pioneering spirit, demonstrating that against tremendous odds and at significant cost, humanity can reach beyond its grasp, fueled by the desire to explore the unknown.

CHAPTER 8

Chapter 8: "Evidence" of Staging and Hollywood Con

The Claims of Hollywood's Role in Staging the Moon Landing

The idea that Hollywood may have had a hand in staging the moon landing has captured the imaginations of conspiracy theorists for decades. This claim draws on the fascination with cinema's power to create convincing illusions and the evolving capabilities of film technology during the 1960s. At the heart of these allegations is the belief that NASA, facing insurmountable challenges in achieving a successful lunar mission, turned to Hollywood experts to produce footage that would simulate the event and bolster public confidence.

The foundation of this theory lies in the impressive strides made by the film industry during that era. By the late 1960s, movie studios had pushed the boundaries of visual effects, producing scenes that, while fantastical, seemed more real than ever before. Films like *2001: A Space Odyssey*, which premiered in 1968, showcased special effects that were groundbreaking, employing innovative techniques such as front projection and meticulously designed models to create a believable portrayal of space travel. Conspiracy theorists argue that if such effects were achievable in a film, NASA could have harnessed similar methods to fabricate the moon landing.

Supporters of the Hollywood theory often point to the need for expertise in simulating the low-gravity environment of the moon. The slow-motion movements of the astronauts, the way dust kicked up by their boots seemed to behave in reduced gravity, and the smooth, sweeping visuals of the lunar landscape are cited as aspects that would have required skilled direction and cinematic knowledge. They argue that without Hollywood's touch, the footage would not have been as convincing as it was.

Rumors surrounding NASA's supposed collaboration with Hollywood gained momentum due to whispers that the U.S. government might have approached filmmakers who were known for their technical prowess. The goal, according to conspiracy theorists, was to create footage that could withstand both public and expert scrutiny, ensuring that America would emerge as the victor in the space race. This collaboration, they claim, would have been essential to crafting scenes that conveyed the triumph of Apollo 11 in a manner believable enough to maintain NASA's credibility and inspire global awe.

Critics of the official moon landing narrative argue that logistical and practical challenges of actually landing on the moon, coupled with Cold War pressures, made faking the footage a more attractive and achievable option than risking an outright failure. In their view, employing Hollywood talent was not only possible but necessary to maintain the illusion. These claims are often supported by anecdotal evidence and conjecture, with theorists asserting that only a select few individuals in NASA and the U.S. government were aware of the plan, thus ensuring secrecy.

While these ideas have a certain allure, they often overlook key historical and logistical details. The film industry's ability to create convincing illusions was impressive, but replicating the totality of the lunar mission, from live transmissions to the movements of astronauts in a vacuum environment, would have required an unparalleled level of coordination and expertise that even Hollywood might not have been able to achieve. Moreover, such an undertaking would have involved numerous individuals who would need to be silenced for decades—a task many researchers argue is far beyond the realm of plausibility.

The concept of Hollywood's involvement in the moon landing hoax has become an enduring part of conspiracy lore, blending the worlds of science, art, and Cold War politics into a single narrative of deceit. This section serves as an entry point into the deeper claims involving notable figures and alleged visual evidence that critics say points to cinematic trickery.

Stanley Kubrick and Alleged Involvement in the Hoax

No discussion about Hollywood's supposed role in the moon landing hoax would be complete without mentioning Stanley Kubrick. The legendary filmmaker behind *2001: A Space Odyssey* has become an almost mythical figure in conspiracy theories related to the moon landing. His groundbreaking work on that film, with its visionary special effects and painstaking attention to realism, serves as the cornerstone of claims that Kubrick could have been recruited by NASA to simulate the Apollo mission.

The theory that Kubrick was involved in staging the moon landing stems from both the timing and the nature of his work. *2001: A Space Odyssey* was released in 1968, just a year before the Apollo 11 mission. The film's portrayal of space travel was so convincing that it immediately became the benchmark for cinematic depictions of space. Proponents of the hoax theory suggest that Kubrick's success in simulating space scenes proved that he possessed the expertise necessary to produce a moon landing that could fool the public. They argue that his meticulous nature and ability to pioneer new techniques made him an ideal candidate for such a task.

One of the most cited pieces of "evidence" comes from supposed confessions and cryptic references found in Kubrick's later work, especially in *The Shining* (1980). Conspiracy theorists point to elements like Danny Torrance's Apollo 11 sweater and the eerie, labyrinthine Overlook Hotel as symbols of Kubrick's alleged guilt and veiled attempt to communicate his involvement. They argue that Kubrick, bound by secrecy, embedded these subtle hints as a form of confession or a nod to those in the know. This interpretation, however, is speculative and heavily reliant on reading intentionality into artistic decisions that may simply be thematic or coincidental.

Additionally, interviews that Kubrick gave throughout his career, as well as alleged deathbed confessions circulated in certain circles, are often presented as further proof. A notable example is a satirical video from 2015 that purported to show Kubrick confessing to faking the moon landing—a clip that, despite being debunked as a hoax, continues to resurface in discussions as supposed evidence. The persistence of such fabrications illustrates how powerful and pervasive these conspiracy theories can become, even in the face of clear refutation.

Experts who have studied Kubrick's life and work, including colleagues and film historians, have refuted these claims. They emphasize that Kubrick was known for being fiercely independent and highly protective of his creative control. The notion that he would engage in a massive government cover-up runs counter to what is known about his personality and artistic integrity. Furthermore, the level of secrecy and coordination required for Kubrick to both direct the alleged fake moon landing and maintain that secret for the rest of his life is implausible to most scholars.

Technically, while Kubrick's work on *2001: A Space Odyssey* did showcase impressive special effects, those effects were not without their limits. The movie employed front projection, miniatures, and carefully crafted sets to create the illusion of space, techniques that, while advanced for cinema, would not have matched the live, continuous broadcasts of Apollo 11. The differences between movie sets and the practical realities of space exploration become apparent when analyzed by experts in photography, video production, and physics. Even the most sophisticated special effects of the time had limitations that would be exposed under the scrutiny that the Apollo footage has faced.

Kubrick's supposed involvement in the moon landing hoax is thus a compelling story, but one that ultimately fails to hold up against factual analysis and historical context. The myth endures, not because of tangible proof, but because of the allure of the filmmaker himself—enigmatic, influential, and capable of creating cinematic masterpieces that blurred the lines between reality and fiction. His name adds a layer of intrigue and drama that conspiracy theories thrive on, feeding into the narrative of a brilliant artist conscripted into a government conspiracy.

In reality, Kubrick's legacy remains that of a man devoted to pushing the boundaries of filmmaking, not one embroiled in secret government operations. The theory that he faked the moon landing serves as a fascinating cultural artifact, more reflective of society's penchant for complex, hidden plots than of any actual involvement by Kubrick in staging one of humanity's most celebrated achievements.

Analyzing Film Techniques and "Proof" in the Apollo Footage

The visual evidence that conspiracy theorists use to claim the Apollo moon landing was staged is often rooted in supposed film techniques of the 1960s. To strengthen their argument, these proponents point to specific anomalies in the footage and photographs from the Apollo missions, arguing that such inconsistencies are telltale signs of studio trickery. This section delves into these claims, presenting both the theories themselves and the counterarguments from experts in film, photography, and science.

One of the most frequently cited pieces of evidence is the peculiar lighting and shadows observed in Apollo photos. Skeptics note that shadows often appear to fall in different directions, as if there were multiple light sources, which they argue would be impossible on the moon, illuminated only by the sun. This, they claim, suggests the use of artificial lighting, such as those found in movie studios. However, photography experts have explained that this effect is actually due to the uneven lunar terrain. The moon's surface is not perfectly flat; it has craters, ridges, and rocks that distort the angles at which light travels and casts shadows. On Earth, this kind of lighting behavior is less noticeable, but in the stark, unfiltered environment of the moon, such contrasts can become pronounced.

Another prominent claim revolves around the iconic image of the American flag appearing to "wave" as if caught in a breeze. The flag's motion, captured in still photographs and video, is cited as proof that the scene was filmed on Earth, where air currents would naturally cause such movement. In response, scientists point out that the flag moved only when astronauts were handling it or had recently planted it into the lunar soil. This movement persisted momentarily due to the flag's stiff, flexible rod and the lack of atmospheric resistance on the moon. Unlike on Earth, there is no air to dampen motion, so any disturbance—whether it's the astronauts' hands or the act of planting the flag—would cause the fabric to oscillate for a longer duration.

Conspiracy theorists also highlight visual artifacts in the background of the lunar footage, such as what they claim are "set boundaries" or repetitive rock formations that suggest painted backdrops or soundstage settings. They argue that in some photographs, specific objects like rocks appear more than once, implying the reuse of props. However, experts in photography and geology argue that these repetitions are due to optical illusions and perspective distortions. The harsh light on the moon and the rudimentary nature of camera technology in the 1960s can create unexpected artifacts that, to an untrained eye, might resemble deliberate staging. Furthermore, modern image analysis has shown that the backgrounds, though similar in appearance due to the barren nature of the lunar surface, are distinct when carefully examined.

The final point in these arguments deals with the transparency and quality of the broadcast itself. Critics assert that broadcasting live video from the moon in 1969 would have been technically impossible due to the supposed limitations of existing communication technology. Yet, engineers and space historians have detailed how NASA used a system of relay stations and radio technology that, while cutting-edge for the time, was entirely feasible. The signal transmission from the moon was encoded, sent back to Earth, and then distributed for public broadcast, explaining any perceived irregularities in the footage's quality.

In reality, many of these alleged discrepancies can be explained through a combination of the unique lunar environment and the limitations of the technology available at the time. Photography and space science experts have analyzed the Apollo footage in detail, demonstrating that features such as shadows, lighting, and camera artifacts align with what would be expected on the moon. The insistence that film techniques were involved persists, however, because these visual "clues" play into the narrative of a well-orchestrated cover-up, satisfying the human penchant for mystery and suspicion.

Ultimately, while it is true that cinema in the 1960s had made significant strides, producing footage on par with the extensive documentation of the Apollo missions would have posed a monumental challenge even for Hollywood. The live broadcast, detailed telemetry data, and thousands of photos, combined with the corroborative testimonies of engineers, astronauts, and scientists, provide substantial evidence that the missions were real. The analysis of these supposed signs of staging underscores that while anomalies may seem suspicious at first glance, they often have straightforward and scientifically sound explanations.

CHAPTER 9

Chapter 9: Why Haven't We Returned? A Lingering Qu

The Hoax Theorists' Perspective on the Absence of Manned Missions

One of the most persistent questions in discussions of the Apollo moon landings is why humanity has not returned to the moon since those historic missions. For conspiracy theorists, this gap serves as a cornerstone of their belief that the moon landings were faked. They argue that if NASA truly succeeded in sending astronauts to the moon six times between 1969 and 1972, it would be reasonable to expect follow-up missions in the subsequent decades. The absence of such missions, they contend, is indicative of a cover-up—one designed to prevent exposure of the supposed deception.

Conspiracy proponents often highlight the sheer audacity of the Apollo missions. To them, the sudden and impressive leap in space travel capabilities during the 1960s, followed by a complete cessation of lunar exploration, appears suspicious. They argue that NASA, under immense pressure from Cold War politics, had the motivation to stage the moon landings to demonstrate technological supremacy over the Soviet Union. By faking the landings, they claim, the U.S. could achieve a political victory without the risk or expense of an actual lunar mission. In this context, the fact that no humans have returned to the moon is presented as proof that the U.S. government and NASA could not replicate the feat without revealing the alleged hoax.

Bill Kaysing, one of the earliest figures to voice doubts about the Apollo missions, emphasized this point in his work. He argued that the technology supposedly available in the 1960s would not have been sophisticated enough to complete a moon mission and return safely. According to Kaysing and those who share his perspective, the reason there have been no further manned moon landings is because it was never achieved in the first place. The hoax theorists posit that NASA avoided organizing new missions to prevent scrutiny that could expose inconsistencies in their original claims.

Further fueling these theories are references to official NASA statements and reports that underscore the difficulty of returning to the moon. In the decades following the Apollo program, NASA faced significant budgetary and logistical challenges. To conspiracy theorists, these explanations are insufficient and viewed as convenient excuses. They argue that if the technology and knowledge to reach the moon existed in the late 1960s, it should have only improved over time, making repeat missions easier and more frequent. This supposed contradiction is cited as evidence of a calculated decision by NASA to avoid further manned missions and thus keep the alleged ruse intact.

In addition to the technological doubts, proponents point to statements made by astronauts and NASA officials that they interpret as cryptic or revealing. Selective quotes are often taken out of context to bolster claims that NASA insiders are aware of the hoax but unable to speak openly due to government pressure. These interpretations are largely speculative and heavily criticized by experts, but they persist within conspiracy circles as tantalizing hints of hidden truths.

Overall, the lack of additional manned moon missions remains a central talking point for those who question the authenticity of the Apollo program. To these theorists, the gap between 1972 and the present day is not a simple case of shifting political and financial priorities; it is seen as the very evidence needed to validate their belief that the moon landings were an elaborate act of deception.

The Practical and Political Challenges Post-Apollo

While conspiracy theorists see the lack of return missions as suspicious, historians and space experts point to a variety of practical and political factors that shaped the trajectory of lunar exploration after Apollo. To understand why no humans have set foot on the moon since 1972, one must examine the geopolitical landscape, funding challenges, and shifting priorities that influenced NASA and the broader context of American space policy.

The Apollo program was born from the intense rivalry of the Cold War, particularly the space race between the United States and the Soviet Union. When President John F. Kennedy announced in 1961 that America would commit to landing a man on the moon and returning him safely to Earth, it was as much a strategic move as it was a scientific one. The Apollo moon landings, culminating in Neil Armstrong's historic step in 1969, achieved that political objective with unparalleled success. The United States demonstrated its technological prowess, capturing the world's imagination and solidifying its position as the leader in space exploration.

However, once the primary goal of beating the Soviet Union to the moon was accomplished, the impetus for continued moon missions waned. The U.S. government faced growing economic challenges, including the costs of the Vietnam War and domestic social programs. These pressures led to significant budget cuts for NASA in the 1970s. The Apollo program, which had been extraordinarily expensive, consuming billions of dollars, was seen as no longer justifiable. The practical and political reality was that American taxpayers and lawmakers began to question the necessity of pouring additional funds into moon missions that no longer served a clear geopolitical purpose.

The cancellation of Apollo missions 18, 19, and 20 serves as a testament to this shift in priorities. Although NASA had plans for further lunar exploration, these missions were scrapped in 1970 due to budgetary constraints. This decision marked a turning point: NASA's focus moved away from sending humans back to the moon and toward more economically feasible projects. The development of the Space Shuttle program, which promised reusable spacecraft that could support a range of missions, represented a strategic pivot. The shuttle was designed for lower Earth orbit tasks, such as satellite deployment, scientific research, and the assembly of the International Space Station (ISS), making it a more practical investment for the agency and the country.

Another significant challenge lay in the logistics of a moon return mission. While NASA had succeeded in its initial moon landings, the infrastructure for such complex missions was dismantled or repurposed after Apollo. Key technologies used in the 1960s, like the Saturn V rocket—the only vehicle powerful enough to carry humans beyond Earth's orbit—were retired, and its production facilities were decommissioned. Rebuilding that capability, particularly after decades of focusing on other types of space missions, would require immense time, effort, and financial resources. This absence of a readily available launch system capable of supporting manned moon missions became a practical barrier that contributed to the lack of return expeditions.

Additionally, advancements in robotic technology during the late 20th and early 21st centuries further reduced the perceived need for human lunar missions. Unmanned probes and rovers could conduct many types of scientific experiments without the risks and costs associated with human spaceflight. These automated explorers became the preferred method for gathering data from celestial bodies, allowing NASA and other space agencies to continue space exploration while avoiding the complex logistics of manned missions.

In sum, the gap in human moon missions post-Apollo can be attributed to a confluence of factors: shifting political priorities, budget limitations, and logistical challenges. Unlike the conspiracy theorists' belief in a cover-up, the historical record suggests that these practical considerations, rather than any hidden agenda, explain why NASA and other space agencies turned their attention elsewhere. The reality of space exploration after Apollo reflects the complexities of balancing ambition, politics, and economic realities in an ever-evolving world.

The Current State of Lunar Exploration and Future Plans

Despite the long hiatus in human moon landings, recent developments have renewed interest in returning to the lunar surface. Modern space agencies, led by NASA and joined by international and private partners, are working on missions that could send astronauts back to the moon for the first time since 1972. This resurgence in lunar exploration challenges the notion that NASA's silence since Apollo was rooted in a cover-up and points instead to the shifting realities of technological advancements, public interest, and strategic priorities.

The Artemis program, launched by NASA in the 2010s, stands at the forefront of current plans for returning humans to the moon. Named after the twin sister of Apollo in Greek mythology, Artemis represents a new era of lunar exploration with ambitious goals: to land "the first woman and the next man" on the moon and to establish a sustainable human presence. Artemis aims not only to revisit past achievements but to surpass them by exploring new regions of the moon, particularly its south pole, which is believed to contain water ice—a crucial resource for future space missions.

The focus on sustainability marks a stark difference from the original Apollo missions, which were designed as short-term excursions. Artemis, in contrast, seeks to create a foundation for long-term exploration, setting the stage for potential missions to Mars and beyond. The program's advancements include partnerships with private space companies such as SpaceX and Blue Origin, which are developing lunar landers and reusable space vehicles that would have been inconceivable during the Apollo era. These partnerships reflect a broader trend of leveraging private sector innovation to achieve governmental space goals, showcasing how space exploration has evolved from a Cold War competition to a more collaborative and economically diversified endeavor.

Conspiracy theorists have struggled to align their narratives with these modern developments. The argument that NASA avoided returning to the moon to cover up a hoax becomes less convincing in the face of new technology, transparent collaborations, and the clear documentation of international participation. If the moon landings had been staged, critics argue, it would be extraordinarily difficult to maintain such a ruse with the numerous private and public players now involved in space exploration. Moreover, many countries and space agencies, including the European Space Agency (ESA), Russia's Roscosmos, and China's CNSA, have their own plans for lunar missions, adding layers of global scrutiny and independent verification that did not exist in the 1960s.

The advancement of space technology also dispels some of the arguments made by hoax proponents about technological limitations. NASA's renewed commitment to moon missions has been driven by progress in rocket engineering, computer systems, and robotics. The upcoming Artemis missions will rely on the Space Launch System (SLS), the most powerful rocket ever built, paired with the Orion spacecraft designed for deep-space travel. These modern engineering feats address the logistical and technical challenges that conspiracy theorists have long cited as evidence against the authenticity of the Apollo missions.

Furthermore, public engagement has shifted. In the decades after Apollo, interest in moon exploration waned, partly because of the perception that the scientific yield did not justify the expense. Today, however, there is renewed enthusiasm, fueled by a new generation inspired by technological achievements and a vision for humanity's role in space. The moon is now viewed as a stepping stone toward greater interplanetary exploration, such as Mars missions. This contemporary rationale contrasts with the Cold War urgency of Apollo and suggests that the absence of manned missions for so long was less about hiding a hoax and more about navigating the evolving priorities of space exploration.

In essence, the reason humans have not returned to the moon until now lies in the confluence of past economic constraints, technological redevelopments, and shifts in geopolitical focus. With Artemis and other international efforts moving forward, the question "Why haven't we returned?" is poised to become obsolete. These new missions highlight that the story of moon exploration is not one of hidden truths but of renewed ambition, strategic evolution, and the human drive to push beyond previous frontiers.

CHAPTER 10

Chapter 10: Witnesses and Whistleblowers – Missing

The Alleged Whistleblowers and Their Stories

Throughout the history of the moon landing hoax theories, one recurring theme has been the suggestion that certain individuals with inside knowledge of NASA's Apollo missions knew the truth and attempted to share it, but were ignored, discredited, or silenced. Central to these claims are figures who have come forward with accounts that suggest discrepancies in the official narrative or outright accusations of a cover-up. Although these stories vary widely in detail and credibility, they have added fuel to the fire of conspiracy theories surrounding the Apollo missions.

One of the earliest and most well-known whistleblowers cited by moon landing hoax proponents is Bill Kaysing. Kaysing was an employee at Rocketdyne, the company that manufactured the engines used in the Saturn V rocket, and he later became one of the most vocal critics of the moon landings. In his self-published book *We Never Went to the Moon: America's Thirty Billion Dollar Swindle*, Kaysing outlined his belief that NASA's moon landings were fabricated to secure a Cold War victory over the Soviet Union. He claimed that inconsistencies in the technology of the 1960s, coupled with high risks and limited chances of success, supported the idea of a staged event. While Kaysing's expertise in rocket propulsion was limited and his arguments were often dismissed as speculative or flawed by experts, his work laid the groundwork for later hoax theories.

Kaysing's assertions were bolstered by anecdotal accounts and selective interpretations of NASA's conduct. For example, he pointed to the behavior of certain Apollo astronauts, such as their supposed discomfort or reluctance during press conferences, as evidence that they were part of an elaborate ruse. Kaysing's arguments, though lacking in empirical support, resonated with audiences who were already inclined to distrust government narratives in the post-Vietnam War and Watergate era.

Other alleged insiders have also surfaced with varying degrees of credibility. Some conspiracy theorists have pointed to Stanley Kubrick, the famed director of *2001: A Space Odyssey*, as a potential whistleblower. They argue that Kubrick's mastery of film techniques made him an ideal candidate to help NASA create convincing footage of the moon landings. This theory gained traction through misinterpretations of his later work, particularly *The Shining*, which some believe contains hidden messages alluding to his involvement. Supporters of this theory claim Kubrick embedded clues as a way of subtly confessing his role while maintaining plausible deniability. However, these interpretations are typically speculative and based on symbolic readings that lack concrete evidence.

The idea of whistleblowers has been reinforced by stories about unexplained deaths or accidents that allegedly befell those who knew too much. Proponents of the hoax theory sometimes cite the deaths of NASA personnel in the 1960s and 1970s as suspicious, hinting that they were orchestrated to prevent potential leaks. These claims often fail under scrutiny, as many of the deaths can be attributed to documented accidents or natural causes unrelated to any conspiracy. Yet the very existence of such stories adds an air of mystery and feeds the belief that individuals who could reveal the truth were silenced before they could do so.

In conclusion, while figures like Bill Kaysing have become synonymous with the moon landing hoax narrative, their credibility and the strength of their arguments remain widely disputed. The lack of compelling evidence from any reputed insider lends more weight to the official account than to theories of a massive, secret cover-up. Nonetheless, the stories of alleged whistleblowers continue to captivate a segment of the public, illustrating how suspicion and the allure of hidden knowledge can perpetuate even the most improbable narratives.

The Challenge of Keeping Thousands Silent

One of the most compelling counterarguments to the moon landing hoax theory is the sheer number of people who would have needed to remain silent for such a cover-up to succeed. The Apollo program was an unprecedented technological feat involving hundreds of thousands of scientists, engineers, technicians, and support personnel across the United States. By conservative estimates, approximately 400,000 individuals contributed directly to the development, execution, and support of the Apollo missions. The notion that such a large number of people could collectively maintain silence about a monumental hoax for decades challenges the very nature of human behavior and organizational secrecy.

Human nature is inherently prone to leaks and disclosures, particularly when large numbers of people are involved in a project of such magnitude. Secrets rarely stay buried when so many individuals share knowledge of them, especially over long periods. Whistleblowers from within major organizations or government agencies have come forward to expose lesser scandals with far fewer participants. This raises a critical question: if the moon landings were indeed faked, why hasn't a credible whistleblower emerged with definitive proof?

Historical examples underscore how difficult it is to maintain silence within large-scale operations. Consider the Pentagon Papers, leaked by Daniel Ellsberg, which exposed secret U.S. military activities during the Vietnam War. Unlike the hypothetical moon hoax, the Pentagon Papers involved a limited circle of officials, yet still, the truth came out. Similarly, Edward Snowden's revelations about the National Security Agency (NSA) demonstrated that even highly classified operations involving national security can be compromised by a single determined individual. These examples highlight that if there were substantive evidence of a fabricated moon landing, someone, at some point over the past five decades, would likely have come forward with more than mere conjecture.

Proponents of the moon landing hoax theory argue that potential whistleblowers have remained silent due to fear of retaliation or the perceived power of the U.S. government to suppress dissent. They suggest that individuals who knew the truth might have been silenced through intimidation, threats to their families, or even elimination. While this notion taps into broader themes of governmental overreach and conspiracy, it fails to account for the diversity of people involved in Apollo's workforce. NASA contractors, scientists from universities, and engineers from private aerospace companies were spread across different locations and had varying degrees of connection to the government. The idea that all these people could be uniformly controlled or coerced without a single substantial leak strains credibility.

The nature of whistleblowing itself is a compelling counterpoint to the conspiracy. People who risk coming forward with the truth are often driven by a profound sense of justice or moral duty. This motivation typically becomes stronger over time, as potential whistleblowers distance themselves from the influence of their former organizations. Given that over 50 years have passed since the last Apollo mission, enough time has elapsed for many potential insiders to feel safe revealing the truth if there were any. Yet, in reality, we find an overwhelming silence on this front—not of suppressed voices, but of consistent support from those who worked on Apollo. Former NASA employees, now retired and with little to lose, still speak about the program with pride and without hesitation, reinforcing the authenticity of their achievements.

Finally, the hoax theory fails to explain why there would be no credible defectors from within the Soviet Union or other rival nations during the height of the Cold War. The Soviet space program, which monitored NASA's progress with great interest and competitive urgency, would have had every incentive to expose a fraudulent moon landing by the United States. If the Soviets had detected evidence of fabrication, they would likely have seized the opportunity to discredit their Cold War adversary and claim victory in the space race. The absence of such a revelation from any foreign power provides a powerful, often overlooked, argument against the possibility of a hoax.

In conclusion, while the idea of whistleblowers being "muzzled" may seem plausible on the surface, the practical realities of human behavior, organizational dynamics, and historical precedent tell a different story. The sheer scale of the Apollo program and the diversity of its participants make the sustained silence required for a hoax almost inconceivable.

Theories on Why Whistleblowers Would or Wouldn't Speak Out

Conspiracy theorists have long speculated why, if the moon landing was indeed faked, there has not been an influx of whistleblowers exposing the truth. The notion of a wide-reaching conspiracy invites questions about the motivations and constraints that would keep so many individuals silent for so long. While some argue that potential whistleblowers may have feared severe consequences, others point to more subtle psychological and social dynamics that could explain their silence.

One popular explanation among hoax proponents is the idea that whistleblowers were deterred by the potential repercussions of speaking out. The U.S. government, with its vast resources and reach, is often depicted in conspiracy circles as an omnipotent force capable of suppressing dissent through coercion or violence. According to this theory, individuals who might have known about a potential hoax faced the threat of ruined careers, public discrediting, or even more drastic measures. This fear of reprisal, it is argued, could have silenced insiders, particularly those with families to protect or financial stability at stake. Supporters of this view often cite mysterious deaths or accidents involving NASA personnel, suggesting that these were more than coincidences and served as warnings to others.

However, while these theories are dramatic and compelling, they lack substantial evidence. Many of the so-called mysterious deaths have reasonable, documented explanations, such as accidents unrelated to space missions or natural causes. More importantly, for such deterrence to work across decades and involve hundreds of thousands of people—many of whom were not high-level officials but regular employees, contractors, and researchers—strains the boundaries of plausibility. Keeping all these individuals silent through fear alone would require an unprecedented level of coordination and secrecy that exceeds known cases of cover-ups in history.

An alternative theory centers on the idea of loyalty and cognitive dissonance. Psychological studies have shown that individuals who invest significant time and energy into a project often develop a strong sense of loyalty and commitment, making it difficult for them to acknowledge that their work may have been in service of something deceptive. For those who worked on the Apollo program, the emotional and professional investment was immense. They were driven not only by the goal of landing a man on the moon but also by the larger narrative of American ingenuity and Cold War victory. For many, accepting that the work they devoted their lives to was part of a hoax would be deeply unsettling. The very notion could trigger cognitive dissonance, a psychological phenomenon where conflicting beliefs cause discomfort, leading people to rationalize or ignore contradictory information to maintain their sense of integrity and purpose.

Social and professional bonds may have further contributed to this silence. The Apollo program was a collaborative effort that united a diverse array of professionals from scientists and engineers to manufacturers and contractors. It fostered a culture of shared achievement and pride. In such an environment, even if someone harbored doubts or uncovered inconsistencies, they would be less inclined to voice them, particularly if it meant betraying colleagues and potentially dismantling the collective legacy. The camaraderie formed during those years would make whistleblowing not just a professional risk, but a deeply personal act of defiance.

On the flip side, one must consider why, if a moon landing hoax existed, someone would choose to speak out after decades of silence. Hoax proponents might argue that with the passage of time, potential whistleblowers would feel safer revealing the truth, either because they had retired from their roles or because they felt a moral obligation to set the record straight as they aged. Yet, this scenario has not materialized. No credible individuals with verifiable insider knowledge have come forward with evidence that the moon landing was staged. The few who have made claims either lack substantial proof or have been easily debunked by experts and historians.

The enduring silence of both American and international experts further reinforces the moon landings' legitimacy. During the Cold War, the Soviet Union would have eagerly exposed any American fraud to undermine the U.S. on the world stage. Instead, Soviet scientists, who tracked the Apollo missions independently, acknowledged the achievements as genuine. The absence of defectors or whistleblowers from NASA's extensive workforce, combined with the validation from independent foreign observers, suggests that there was no hoax to reveal.

In conclusion, the silence surrounding potential whistleblowers from the Apollo program can be explained by a mix of loyalty, professional pride, cognitive dissonance, and the high standards of evidence required for credible claims. The speculative theories of fear or intimidation are far less convincing than the collective testimony of those involved and the global acknowledgment of the missions' success.

CHAPTER 11

Chapter 11: Rebuttals from NASA and Scientific Com

NASA's Official Stance on Hoax Theories

Since the first whispers of moon landing conspiracy theories surfaced, NASA has consistently maintained a clear and firm stance: the Apollo moon landings were real. These missions were the culmination of years of dedicated work by scientists, engineers, and astronauts who pushed the boundaries of human knowledge and capability. While conspiracy theories can be captivating, NASA has frequently stepped forward to address and dismantle the arguments made by hoax proponents, relying on decades of documented evidence, technical data, and firsthand accounts to reinforce the authenticity of the missions.

NASA's public rebuttals began almost as soon as the conspiracy theories did. In the early 1970s, shortly after the Apollo missions concluded, various books and articles claimed that the moon landings were staged. In response, NASA published detailed reports and made mission archives available for scrutiny. One notable response came in the form of the agency's comprehensive Q&A segments, where NASA officials addressed questions about the alleged anomalies pointed out by skeptics. For example, the agency explained why the American flag appeared to be "waving" on the moon's surface—a common claim by hoax advocates. The flag's motion was due to its aluminum support rod and the residual movement when astronauts planted it into the lunar soil, not the presence of wind, which does not exist on the moon.

NASA has also leveraged the testimonies of Apollo astronauts to reinforce their position. Buzz Aldrin, the second man to walk on the moon, has been outspoken about the reality of the missions. In interviews, Aldrin has recounted his experiences on the lunar surface in vivid detail, describing not only the technical aspects but the profound personal impact of stepping onto an alien world. In one widely publicized incident, Aldrin famously confronted a conspiracy theorist who accused him of lying about his moonwalk, demonstrating the frustration felt by many who were part of the missions and have spent years defending their truth.

In addition to astronaut testimonials, NASA's outreach efforts include sharing a wealth of primary source materials. The agency has made photographs, mission transcripts, and technical blueprints publicly available, allowing independent researchers and space enthusiasts to verify the details themselves. These archives provide an unparalleled level of transparency and have been utilized by historians, scientists, and even laypeople to refute the claims that the moon landings were an elaborate hoax. The sheer volume of verifiable data—from communications between the lunar module and Mission Control to the precise telemetry records—offers evidence that would be difficult, if not impossible, to fabricate on such a grand scale.

NASA has also leaned on the broader scientific community to help address technical claims that suggest the landings were staged.

The space agency frequently points to the hundreds of academic and research institutions that contributed to the Apollo missions, emphasizing that a conspiracy of this magnitude would require complicity from a wide range of scientists, engineers, and experts beyond just those employed by NASA. The collaboration with these external entities lends credibility to the Apollo program's authenticity, showcasing it as a multinational and multidisciplinary effort.

In conclusion, NASA's official rebuttals against moon landing conspiracy theories are rooted in transparency, technical data, and personal testimonies from those who witnessed and made history. By providing open access to mission records, detailed explanations of alleged anomalies, and statements from astronauts themselves, NASA continues to uphold the legacy of the Apollo missions as one of humanity's greatest achievements.

Explanations for Alleged Photographic Anomalies

One of the most frequently cited pieces of "evidence" by moon landing hoax proponents lies in the photographs and video footage taken during the Apollo missions. From the odd angles of shadows to the seemingly impossible lighting conditions and the conspicuous absence of stars in the lunar sky, these claims have captured the imaginations of skeptics for decades. However, experts in photography, physics, and space science have thoroughly analyzed these supposed anomalies, providing explanations grounded in the realities of lunar conditions and camera technology of the 1960s.

A prominent point of contention involves the shadows visible in the Apollo photos. Skeptics argue that the presence of shadows at different angles suggests multiple light sources, implying that the footage was staged on a set illuminated by artificial lights. However, photography and lighting experts debunk this by explaining the unique way light behaves on the moon. Unlike on Earth, where light diffuses through the atmosphere and softens shadows, the moon lacks an atmosphere to scatter sunlight. As a result, shadows on the lunar surface are stark and high-contrast. Moreover, the uneven, rocky terrain of the moon contributes to the appearance of shadows pointing in different directions. The moon's uneven ground causes light to scatter in ways that create complex shadow patterns, making it appear as though there are multiple light sources when there is only one: the sun.

Another common argument is that the stars are conspicuously absent from the lunar sky in all Apollo mission photographs. Conspiracy theorists claim that the omission of stars was intentional, as accurately recreating star positions on a soundstage would have been too difficult. In reality, the absence of stars is a straightforward consequence of photographic exposure. The cameras used during the Apollo missions were set to capture the bright, sunlit surface of the moon and the astronauts' activities. The lunar surface reflects sunlight intensely, requiring camera settings with short exposure times and narrow apertures to avoid overexposure. These settings effectively wash out any dim light, including the faint starlight, rendering it invisible in the images. Photography experts and astronauts have confirmed that seeing stars would have required adjusting the camera settings to capture low-light conditions, which would have overexposed the moon's bright surface.

The famous "waving flag" incident is another anomaly often cited by hoax theorists. In footage from the Apollo missions, the American flag appears to flutter as if caught in a breeze—impossible in the moon's vacuum. But this effect is easily explained. The flag was specially designed with a horizontal rod at the top to keep it extended, giving it the appearance of waving when disturbed. When astronauts planted the flag into the ground, the twisting motion caused it to ripple momentarily. Without air resistance to dampen the motion, the flag continued to oscillate for longer than it would on Earth. This characteristic of motion in a vacuum—where there is no medium to slow down movement—created an illusion of fluttering. Physicists have pointed out that the behavior of the flag aligns perfectly with what one would expect in a zero-atmosphere environment.

Expert analyses of these photographs extend beyond theoretical explanations. Independent researchers and photographic experts have recreated scenarios that mimic the conditions of the Apollo missions. Using similar cameras and materials under controlled conditions that replicate lunar lighting, they have demonstrated that the supposed anomalies can be recreated on Earth under specific, realistic parameters. These reproductions reinforce the argument that the Apollo mission photos and videos are consistent with what one would expect from images taken on the moon.

In essence, the supposed photographic inconsistencies pointed out by moon landing skeptics fall apart under expert scrutiny. The unique environmental conditions of the moon, combined with the limitations and specifics of 1960s camera technology, explain why these images appear the way they do. From the strange angles of shadows to the absence of stars and the "fluttering" flag, each claim has a scientifically sound explanation that supports the authenticity of the Apollo missions.

Addressing Technical and Environmental Counterarguments

One of the key arguments made by moon landing hoax proponents revolves around the technological capabilities of the 1960s. Skeptics often assert that the space suits, the lunar module, and other equipment would have been inadequate to protect astronauts from the harsh realities of space travel, including the extremes of temperature, vacuum pressure, and cosmic radiation. However, extensive documentation from NASA, input from engineers who worked on the Apollo program, and subsequent analyses by space technology experts all highlight that these arguments overlook both the ingenuity and precision engineering behind the missions.

The space suits worn by Apollo astronauts, known as Extravehicular Mobility Units (EMUs), were technological marvels designed specifically to withstand the lunar environment. Each suit was composed of multiple layers that combined materials such as Teflon, Mylar, and Dacron, creating a structure that balanced flexibility, durability, and insulation. Skeptics often argue that these suits could not have protected the astronauts from the extreme temperatures on the moon, which range from scorching highs of about 127°C (260°F) in sunlight to frigid lows of around -173°C (-280°F) in shadow. Yet, NASA engineers devised a solution by incorporating an integrated life-support system that regulated temperature and provided a constant flow of oxygen. This system, along with the suit's reflective outer layer, effectively maintained a stable internal environment, allowing astronauts to perform their tasks on the lunar surface without risk of heat exhaustion or freezing.

The lunar module itself has also been a target of skepticism. Hoax proponents frequently point to its seemingly fragile structure, questioning how such a delicate-looking craft could have landed on and lifted off from the moon. This perception, however, fails to account for the precise engineering that went into designing a vehicle tailored for the moon's conditions. The lunar module was not built to endure atmospheric re-entry or Earth's gravitational forces but to operate within the moon's weaker gravity, which is about one-sixth that of Earth. This reduced gravity allowed for a lighter structure that would be sufficient for both landing and takeoff. Additionally, the module's descent and ascent engines were meticulously calibrated to provide the necessary thrust for a smooth landing and a safe return to lunar orbit. The structural integrity of the lunar module has been validated through numerous tests, surviving real-world conditions during the Apollo missions.

Another argument that conspiracy theorists pose is that the technology required to protect astronauts from cosmic radiation, particularly in space beyond the protective shield of Earth's magnetosphere, was lacking. The most frequently cited obstacle is the Van Allen radiation belts, which encircle the Earth and present significant levels of radiation. Detractors claim that passing through these belts would have exposed astronauts to lethal doses of radiation. However, experts in space physics explain that the Apollo missions were carefully planned to minimize exposure. The trajectory used for the lunar missions passed through the thinnest sections of the radiation belts, and the astronauts spent only a short amount of time transiting through them, which limited their total exposure. The spacecraft itself was shielded sufficiently, as confirmed by scientific studies and dosimetry reports, which showed that the radiation exposure was well within safe levels for the duration of the mission.

Furthermore, the commitment of scientists and engineers to overcoming these challenges was rooted in rigorous planning, years of testing, and a relentless pursuit of reliability. Each phase of the Apollo program, from the Saturn V launch vehicle to the return trajectory and re-entry procedures, was meticulously developed with input from the brightest minds of the era. Independent engineers and aerospace historians today continue to affirm that the technology available in the 1960s, though less advanced by today's standards, was more than capable of achieving the goals of the Apollo program.

In summary, while conspiracy theories may raise questions about the feasibility of 1960s technology handling the rigors of space travel, these doubts do not stand up to scrutiny. The Apollo program's achievements were the result of an unprecedented collaboration of science, engineering, and innovation, all documented in painstaking detail and affirmed by countless experts. From the multi-layered space suits to the efficient engineering of the lunar module and the safe passage through cosmic radiation zones, the technical and environmental barriers cited by skeptics have been effectively addressed by decades of thorough analysis and firsthand accounts.

CHAPTER 12

Chapter 12: The Psychological Appeal of Conspiracy

The Cognitive Underpinnings of Belief in Conspiracies

The allure of conspiracy theories is rooted deep in the way the human brain processes information. Our cognitive systems, evolved to make sense of the world, are both marvelously complex and inherently flawed. One of the most powerful cognitive biases that drive belief in conspiracy theories is **confirmation bias**. This psychological tendency compels individuals to seek out, favor, and interpret information in a way that confirms their existing beliefs. For those inclined to question mainstream narratives, confirmation bias acts as a filter, amplifying evidence that supports the belief that the moon landing was a hoax while dismissing or undervaluing evidence that contradicts it. When a conspiracy theorist finds an article, video, or comment that aligns with their viewpoint, the brain experiences a rush of satisfaction that reinforces the bias, making it even harder to accept new information that challenges this belief.

Closely linked to confirmation bias is the human tendency to engage in **pattern recognition**. This trait is an evolutionary adaptation that helped early humans survive by identifying dangers and opportunities in their environment. Spotting patterns quickly allowed for swift decision-making—essential when faced with ambiguous situations in the wild. However, this same survival mechanism can lead modern minds astray. The moon landing photos, for instance, with their stark shadows and peculiar lighting, provide ample material for pattern recognition gone awry. Individuals scrutinizing these images might spot what they perceive as inconsistencies or hidden signs, interpreting them as "proof" of a staged event. In reality, these "patterns" often result from a misunderstanding of how light behaves in the vacuum of space or the technical limitations of 1960s photographic technology.

Another cognitive factor that fuels conspiracy thinking is **proportionality bias**. This is the belief that significant events must have equally significant causes. The moon landing was one of the most monumental achievements in human history—a feat of such grandeur that some individuals struggle to accept that it could have been accomplished by humans without an equally extraordinary hidden explanation. The sheer scope of the Apollo program, with its massive budget, ambitious goals, and political stakes, seems to demand an extraordinary backstory. For conspiracy theorists, the idea that the event might have been staged resonates more than accepting that a team of thousands of engineers, scientists, and astronauts worked in tandem to make it a reality.

Moreover, the **Dunning-Kruger effect** plays a pivotal role in why people cling to conspiracy theories. This cognitive bias causes individuals to overestimate their knowledge or expertise in areas where they have limited understanding. Those who have not studied aerospace engineering, physics, or space science might misinterpret technical data or photographic evidence due to a lack of specialized knowledge, yet feel confident in their assessments. This overconfidence often fuels the belief that they have uncovered "truths" overlooked by experts, reinforcing their allegiance to the hoax theory.

These cognitive biases are natural and deeply embedded in the way we interpret the world. Understanding them is key to recognizing why so many people are drawn to conspiracy theories like the moon landing hoax. The combination of seeking validation for pre-existing beliefs, identifying supposed patterns, assigning proportional significance to grand events, and overestimating one's own understanding creates a fertile ground for the persistence of such theories. While these biases can make individuals feel as though they are uncovering hidden truths, they often lead to conclusions that are detached from scientific reality and robust evidence.

Sociological and Cultural Influences

The appeal of conspiracy theories extends beyond individual cognitive biases and taps into broader sociological and cultural dynamics. Conspiracy theories, like those surrounding the moon landing, often find fertile ground in societies experiencing high levels of distrust in authority. In the context of the Cold War era, when the Apollo missions took place, public skepticism about government transparency and motives was heightened by a backdrop of espionage, political secrecy, and propaganda. Fast forward to today, and similar waves of distrust persist, amplified by modern communication platforms and the instantaneous spread of information. Understanding why people gravitate toward such theories requires exploring the societal structures and cultural narratives that encourage skepticism and shape beliefs.

A significant factor contributing to the popularity of moon landing hoax theories is **social identity**. People often derive a sense of belonging and validation from aligning with groups that share their beliefs. The internet has made it easier than ever for like-minded individuals to find and reinforce each other's views, creating communities centered around alternative explanations. Within these groups, the idea that they collectively hold a "secret" or "special" knowledge sets them apart from mainstream society. This communal aspect is powerful—it taps into the deep-seated human need for connection and a shared identity. Being part of a group that challenges accepted narratives fosters an "us versus them" mentality, reinforcing the idea that the group is enlightened while the majority are deceived.

The role of **groupthink** further entrenches belief in conspiracy theories. Once an individual is immersed in a group that promotes a particular idea, like the moon landing hoax, dissenting opinions are often discouraged or met with hostility. This pressure to conform solidifies beliefs and discourages critical examination of the evidence that contradicts the group's view. The result is an echo chamber effect, where the same ideas are circulated, validated, and amplified without external scrutiny. This dynamic is particularly potent in the digital age, where social media algorithms can create curated feeds that filter out opposing viewpoints and present only the content that aligns with and reinforces a user's beliefs.

Culturally, conspiracy theories have a long history of flourishing during periods of societal upheaval and uncertainty. The moon landing itself was conducted during the tumultuous 1960s, a decade marked by civil rights struggles, political assassinations, the Vietnam War, and widespread protests. Distrust in government and authority was at an all-time high, and public disillusionment made fertile ground for skepticism about major government achievements. Today, modern parallels exist: economic inequalities, political polarization, and global crises create a similar environment where people are more inclined to question official accounts and seek alternative explanations.

The influence of **media and entertainment** has also shaped the public's perception of what might be possible behind closed doors. Pop culture has portrayed elaborate government cover-ups in movies and TV shows, from espionage thrillers to science fiction epics. The idea of a monumental event like the moon landing being staged fits neatly into these narratives, making it easier for people to imagine how such a hoax could unfold. This crossover between fiction and perceived reality blurs the line and provides a framework in which conspiracy theories can thrive. For some, the familiarity of these themes, portrayed repeatedly in various forms of media, reinforces the belief that such conspiracies are not only possible but plausible.

In essence, sociological and cultural factors work in concert to bolster the allure of moon landing hoax theories. Group identity and collective reinforcement provide a sense of belonging, while societal distrust and the echo chamber effect ensure that skepticism remains strong. Add to this the influence of media shaping perceptions of government capability and duplicity, and the appeal becomes deeply rooted in the cultural consciousness. These factors help explain why, decades after the Apollo missions, some still question the reality of one of humanity's greatest achievements.

The Emotional and Psychological Appeal

The emotional draw of conspiracy theories goes beyond mere intellectual curiosity—it often fulfills deeper psychological needs and provides an anchor in a world that feels unpredictable and complex. For those who believe that the moon landing was staged, the theory offers more than just an alternative historical explanation; it provides a framework for understanding human nature, power, and control.

At the heart of this appeal lies the human desire for certainty, the psychological need to be part of a select group that "knows the truth," and the emotional rewards tied to defying mainstream narratives.

One of the strongest emotional motivators for believing in conspiracy theories is the **desire for control** in a chaotic world. The moon landing, an event that showcased unprecedented technological prowess and human achievement, unfolded against a backdrop of societal upheaval. For many, the 1960s were marked by civil rights struggles, the Vietnam War, and political scandals. In times of turbulence and change, people often look for explanations that fit their view of a world that is less arbitrary and more structured. Conspiracy theories fulfill this need by presenting the idea that monumental events are orchestrated by powerful forces rather than left to random chance or sheer human effort. To believe that the moon landing was a hoax implies that events are controlled by an elite, providing a sense of predictability—even if it comes with a cynical outlook.

Another psychological driver is the **sense of belonging** and the allure of holding exclusive knowledge. Being part of a group that challenges widely accepted beliefs offers a feeling of distinction and intellectual independence. For believers in the moon landing hoax, there is an intrinsic reward in identifying with a community that claims to see beyond the mainstream "deception." This dynamic is strengthened by the perception of being "awake" or more perceptive than others, leading to a sense of superiority and satisfaction. The communal aspect of conspiracy theory groups also reinforces loyalty to the belief system; shared skepticism and mutual validation bolster commitment, making it harder to abandon the theory even when faced with contradictory evidence.

The **emotional rush of discovery** cannot be overlooked when examining why people are drawn to conspiracy theories. Uncovering "hidden truths" can be an exhilarating experience. This rush, which is underpinned by dopamine release in the brain, creates a feedback loop that keeps individuals searching for more clues and evidence. This cycle of discovery is compelling and self-perpetuating: each new piece of "evidence" feeds the belief system and brings a fresh wave of satisfaction, ensuring the commitment to the theory deepens over time.

Challenging mainstream beliefs also taps into the **psychological appeal of rebellion**. For some, embracing a conspiracy theory is an act of defiance against what they perceive as the dominant, controlling structures of society. The moon landing, as a crowning achievement of government and institutional collaboration, represents an easy target for those who are skeptical of official narratives. The idea that the government could orchestrate a massive, elaborate hoax fits well into the broader sentiment of mistrust that conspiracy theorists often feel toward authority figures. This antagonistic stance is empowering; it allows believers to position themselves as courageous dissenters, fighting against the tide of blind acceptance.

Lastly, the emotional payoff of these beliefs is not just rooted in skepticism but in the broader idea of **affirming one's worldview**. If one already harbors beliefs about government deception or corporate greed, the moon landing hoax theory becomes a puzzle piece that fits snugly into a larger picture. This sense of consistency, even if it is built on shaky ground, is comforting. It allows individuals to reinforce their belief system, turning what might have started as curiosity into a key pillar of their understanding of how the world works.

In summary, the psychological appeal of moon landing conspiracy theories is multifaceted, involving a mix of cognitive rewards, emotional satisfaction, and social connection. Theories thrive not just on doubts but on the complex interplay of emotional needs, from a desire for certainty to the thrill of perceived discovery. This blend of factors ensures that conspiracy theories, like the moon landing hoax, maintain their grip on the imaginations and beliefs of many, long after the original event has passed into history.

CHAPTER 13

Chapter 13: The Influence of Media and Pop Culture

The Role of Early Media and Documentaries

The moon landing hoax theory, while popular today, owes its origins to a handful of early voices that questioned NASA's achievements in the years following the Apollo missions. The late 1960s and early 1970s were filled with headlines celebrating the United States' triumph in space, yet within this period of national pride, seeds of doubt were sown by a few who would not be silenced. Central to this initial wave of skepticism was Bill Kaysing, whose 1976 self-published book, *We Never Went to the Moon: America's Thirty Billion Dollar Swindle*, laid the foundation for what would become a lasting conspiracy theory.

Kaysing, who once worked as a technical writer for a contractor to NASA, claimed insider knowledge and portrayed himself as an unlikely whistleblower. His assertions were bold: he suggested that NASA lacked the technology and capability to safely send astronauts to the moon and return them. Kaysing's book, though widely criticized for its lack of technical accuracy and speculative nature, struck a chord with a public already disillusioned by events like the Vietnam War and the Watergate scandal. Trust in government institutions was eroding, and Kaysing's narrative played perfectly into a growing cultural skepticism. The timing of his publication was critical—people were ready to question authority, and his work planted the seed that would later grow into a full-fledged movement.

The reach of early media didn't stop with print. The late 1970s and 1980s saw an increase in televised specials and documentaries that flirted with conspiracy theories. One of the most influential programs was a 1978 television special titled *UFOs: It Has Begun*, which, while primarily focused on extraterrestrial phenomena, subtly reinforced the idea that authorities could hide grand secrets from the public. Television served as an accessible platform for theories that might have otherwise been relegated to the fringes. The visual medium allowed conspiracy theorists to showcase selective photographic evidence, play with shadow and light discrepancies in moon footage, and imply staged scenarios—all compellingly presented to millions of viewers in their own living rooms.

The role of radio during this period should also not be underestimated. Programs that featured talk-show hosts open to alternative theories allowed moon landing skeptics to share their views with audiences looking for alternative explanations. Unlike television, radio could entertain longer discussions, dive deep into supposed inconsistencies, and host debates between skeptics and experts. This approach lent the theory a veneer of credibility and engaged listeners in dialogue that legitimized skepticism as more than mere fringe thinking.

These early media efforts set the stage for what would become a powerful movement. They leveraged a combination of shock value and intrigue, suggesting that the moon landing was not just an event to celebrate but a story to investigate. The public appetite for mysteries and hidden truths was insatiable, and the idea that NASA could have faked such a monumental event took root in the collective consciousness. By the time the internet emerged as a force in the 1990s, the groundwork had been laid. The moon landing hoax theory had evolved from whispers in print and discussions on late-night radio to a widely debated topic with a foothold in mainstream culture.

Movies, Television, and Fiction as Catalysts

The spread of the moon landing hoax theory was significantly bolstered by popular culture's fascination with space exploration, government secrecy, and the fine line between fact and fiction. Among the most influential works contributing to this was the 1978 movie *Capricorn One*, directed by Peter Hyams. While not directly about the moon landings, this film told the gripping story of a staged Mars landing orchestrated by a desperate government trying to protect its reputation. The plot featured astronauts forced to fake the mission in a studio, suggesting that maintaining national prestige justified any means necessary. Audiences watching *Capricorn One* found themselves confronting a scenario that blurred fiction and reality, giving weight to the idea that if a government could fake one thing, why not the moon landing?

The impact of such films was not limited to just entertainment; they reinforced doubts that were already simmering beneath the surface. People who were already skeptical of official accounts could point to *Capricorn One* as an example of plausible deception. The film utilized techniques that resonated with conspiracy theorists: a narrative of high stakes, a focus on media manipulation, and the cold, calculating nature of institutional power. This movie, among others, helped cement the notion that massive secrets could be concealed behind the polished façade of space programs. It painted a picture of a government capable of going to extreme lengths to achieve its goals, making real-life conspiracy theories like the moon landing hoax seem far less implausible.

Television shows of the 1990s and early 2000s further fueled these ideas. *The X-Files*, a series that epitomized the era's obsession with hidden truths and powerful cover-ups, featured storylines where government plots intersected with extraterrestrial and space themes. The show's success lay in its ability to make the unbelievable believable, engaging millions of viewers who started to question more than just alien abductions. This blend of suspense, investigative storytelling, and nuanced character skepticism left viewers open to the possibility that not everything they'd been told by authorities was true. While *The X-Files* never explicitly focused on the moon landing hoax, its emphasis on secrets and conspiracies created fertile ground for related theories to gain traction.

The reach of pop culture didn't stop at film and TV; books and novels also played a crucial role. Works of fiction that suggested government duplicity or secret space missions found a receptive audience among readers eager to explore the shadowy side of space exploration. These stories gave readers not only a thrill but also fodder for the imagination, blending well-researched technical details with speculative plots. Such literature often featured protagonists who challenged official narratives, resonating with real-world conspiracy theorists who viewed themselves as truth-seekers pushing against a tide of mainstream belief.

These fictional portrayals laid a framework that made it easier for real-life theories to flourish. A film like *Capricorn One* or a long-running show like *The X-Files* did more than entertain; they validated a worldview where powerful entities could manipulate facts and orchestrate grand deceptions. This alignment between popular culture and conspiracy theories created an echo chamber where ideas reinforced one another, making it easier for skeptics to believe that NASA could have orchestrated an elaborate hoax. The visual language of these shows and movies—studios that mimic space, menacing government agents, hidden evidence—translated into powerful images that stuck with viewers and readers.

By embedding the possibility of hoaxes and cover-ups within the broader context of entertainment, popular culture did not just reflect public doubt; it amplified and normalized it. As a result, when audiences engaged with content questioning the moon landings, they were not just considering dry facts and figures but recalling the dramatic scenarios they'd seen on screens and read in books. This multimedia reinforcement perpetuated a cycle of skepticism, strengthening the resolve of those who believed in the possibility of deception.

The Internet, Social Media, and Modern Pop Culture

The advent of the internet in the 1990s revolutionized the way people accessed and disseminated information, creating an unprecedented platform for conspiracy theories to thrive. The moon landing hoax theory, which had simmered on the fringes of popular thought for decades, found fertile ground in this new era. Websites, blogs, and online forums became the digital meeting places for skeptics who previously relied on niche publications and late-night radio shows. Suddenly, a movement that once depended on a few influential books and television specials could proliferate at a global scale.

In the early days of the internet, amateur theorists seized the opportunity to share and dissect so-called evidence, from grainy photos allegedly showing lighting anomalies to selective video clips featuring astronauts appearing weightless. Websites devoted exclusively to moon landing skepticism, complete with articles, annotated images, and forum discussions, began to emerge. This virtual content was accessible to anyone, from the mildly curious to the deeply invested. People no longer needed to scour libraries for obscure books or hope to catch a conspiracy-themed TV special; they could engage with like-minded individuals and follow detailed arguments, all from the comfort of home.

As the 2000s progressed, video-sharing platforms like YouTube became the primary means of reaching audiences. Videos could combine visuals, voiceovers, and special effects to build compelling cases against the authenticity of the Apollo missions. Home-grown documentaries critiquing NASA's moon landing footage, sometimes in stark and sensationalized terms, spread quickly. Even more reputable channels and filmmakers occasionally released content that touched on the moon landing debate, sparking renewed public interest. These videos ranged from analytical breakdowns of technical claims—highlighting things like unexplained shadows or alleged inconsistencies in photographs—to dramatic exposés promising to reveal "the truth" behind the Apollo missions.

Social media further amplified these theories by providing instantaneous sharing and community building. Platforms like Facebook and Twitter allowed conspiracies to evolve into trending topics, drawing attention from new audiences and fostering rapid debate. Memes simplified complex ideas, making them easily consumable and shareable, while eye-catching graphics and snippets of moon landing footage could be repurposed to make skepticism both viral and entertaining. The speed at which content could be shared meant that theories were no longer confined to small circles; they became part of mainstream internet culture.

Ironically, the democratization of information through the internet led not only to greater spread but also to a paradoxical amplification of doubt. The availability of high-quality analysis and well-researched rebuttals did little to deter those committed to believing in the hoax. Instead, skeptics could cherry-pick data that suited their views, dismissing expert debunking as another arm of government or establishment propaganda. Confirmation bias flourished, as algorithms rewarded engagement by showing users more of the content they interacted with—whether it supported the moon landing or argued against it.

Meanwhile, modern pop culture adapted, continuing to contribute to the dialogue around the moon landings through new lenses. Documentaries released by streaming platforms explored both sides of the argument, maintaining a veneer of neutrality while ensuring that sensationalism kept viewers hooked. The resurgence of space-themed movies and television series kept public interest in space exploration alive, and conspiracy theories often came up in discussions surrounding these releases. Public figures and influencers, wielding significant reach on YouTube and social media, sometimes lent their voices to moon landing debates, either by questioning official accounts or debunking the theories.

By the 2010s and beyond, the moon landing hoax had become a well-entrenched part of conspiracy culture, adapted to modern communication tools and interwoven with broader themes of distrust in authority and questioning official narratives. The digital age had taken what started as a provocative claim by a few and transformed it into a widespread, constantly evolving conversation—a testament to the power of modern media in shaping and sustaining belief. In this new landscape, truth and skepticism dance a complicated dance, with the stage set by algorithms and platforms that thrive on curiosity, doubt, and endless sharing.

CHAPTER 14

Chapter 14: Surveying Public Opinion – Belief in t

Historical Trends in Public Belief

The moon landing on July 20, 1969, marked an extraordinary milestone in human history, watched by an estimated 600 million people worldwide. For most, it was a moment of triumph—proof that humankind could reach beyond Earth and set foot on another world. Yet even in the early days of Apollo 11's success, whispers of doubt began to surface. The moon landing hoax theory, now a well-known part of conspiracy culture, had roots in an era when skepticism towards institutions was burgeoning. Understanding how public belief in the hoax evolved over time requires a look back at history's shifting tides of trust and the emergence of media channels that gave these theories a voice.

In the late 1960s and early 1970s, belief in a moon landing conspiracy was largely confined to fringe groups. Bill Kaysing's self-published book, *We Never Went to the Moon: America's Thirty Billion Dollar Swindle* (1976), played a critical role in seeding public doubt. Kaysing, a former technical writer for Rocketdyne—a company that had worked on NASA's engines—claimed insider knowledge about NASA's alleged deception. His book offered up what he considered evidence: inconsistent shadows in photographs, the absence of stars in lunar sky images, and the now-famous "waving flag" anomaly. Though his arguments lacked rigorous scientific backing, the book resonated with those predisposed to distrust governmental power and spurred a wider conversation.

The skepticism gained subtle momentum as the 1970s unfolded, bolstered by a period rife with political scandals that sowed seeds of distrust in the American public. The Watergate scandal, in particular, left a lasting impression that the U.S. government was capable of complex cover-ups. This era of diminished faith in authority coincided with an increase in people questioning the moon landings, as the newly cynical public became more open to the possibility of mass deception.

By the 1980s, belief in the moon hoax theory was still a niche topic, perpetuated through newsletters, underground publications, and occasional radio broadcasts. However, as the conspiracy theories intersected with a growing counterculture movement that celebrated challenging mainstream narratives, they found pockets of dedicated believers. This underground period laid the foundation for what would later become a mainstream curiosity.

The 1990s marked a significant turning point with the dawn of the internet and the rapid spread of information. Early web users discovered a platform where voices questioning conventional narratives could flourish. Small conspiracy communities that were once limited by geography or access to specialized literature found a digital space to connect and share theories. Websites dedicated to discussing government cover-ups, including the moon landing, emerged and garnered a loyal following. Suddenly, skepticism that had simmered on the fringes of society had a platform that could reach millions.

Polls taken during this period indicated that while belief in the hoax remained a minority view, it was gaining attention. Surveys from the late 1990s and early 2000s showed that between 5% and 10% of Americans believed the moon landings were faked. While this number was still relatively small, it was indicative of an enduring, if not growing, culture of doubt. The combination of historical skepticism fueled by political mistrust and the internet's rise created fertile ground for the moon landing hoax theory to transition from obscure to widely recognized.

Demographic and Regional Analysis of Hoax Belief

The moon landing hoax theory is not confined to any single demographic or region; its presence can be found across a wide spectrum of individuals. However, studies and surveys reveal that belief in the moon landing hoax varies significantly based on factors such as age, education level, and geographic location. Understanding these variations provides insight into the broader appeal of conspiracy theories and the social contexts that sustain them.

In recent decades, comprehensive surveys have attempted to map out who believes in the moon landing hoax. Research shows that belief is more prevalent among younger generations who did not experience the Apollo missions firsthand. A 2019 survey conducted by YouGov found that about 11% of Americans aged 18-34 believed that the moon landings were staged, compared to only 3% of those over 55. This generational divide can be attributed to several factors, including the distance from the historical event itself and changing attitudes toward governmental and institutional trust. Older generations, who grew up in an era when national pride was closely tied to the space race, may find the idea of a hoax less plausible and more offensive to their collective memory.

Education also plays a significant role in belief in the moon landing hoax. People with higher levels of formal education, particularly in the sciences, are less likely to subscribe to the theory. This correlation suggests that scientific literacy and exposure to rigorous analytical thinking help counteract susceptibility to conspiracy theories. Conversely, individuals with limited exposure to education that emphasizes critical thinking or scientific principles are more likely to entertain the idea that the moon landing was fabricated. This is not to say that belief in the hoax is limited to the uneducated; it can be found in all educational backgrounds, but it is more concentrated where educational opportunities have been fewer or less robust.

Geographically, the United States remains the epicenter of the moon landing hoax belief, which is unsurprising given the central role NASA and the U.S. government play in the narrative. However, belief in the hoax is not confined to American borders. Surveys and anecdotal reports indicate that skepticism about the moon landings can be found worldwide, particularly in countries where anti-American sentiment or distrust of Western media runs high. In some parts of the world, belief in the moon hoax is tied to broader political ideologies or historical grievances, serving as a convenient metaphor for Western deception.

In Europe, attitudes vary. While the majority of the public in countries like Germany, the United Kingdom, and France generally accept the Apollo missions as legitimate, pockets of moon landing skepticism persist, fueled by both nationalist biases and the reach of global conspiracy theories through digital media. In Russia, home to the former Soviet Union and NASA's primary rival during the space race, a curious mix of respect for space achievements and skepticism toward American accomplishments creates an interesting landscape. While many Russians hold their country's space program, including pioneering figures like Yuri Gagarin, in high regard, a notable segment of the population entertains doubts about NASA's success. This skepticism is sometimes bolstered by state-sponsored media or nationalist rhetoric that underscores competition with the West.

The connection between belief in the moon landing hoax and trust in media cannot be understated. Polls show that individuals who already view mainstream news outlets with suspicion are more likely to believe in conspiracy theories, including the moon landing hoax. This distrust can be compounded by political beliefs; those with populist or anti-establishment leanings are particularly susceptible. The overlap between political ideology and conspiracy theory belief becomes apparent when examining public opinion in regions where populist or authoritarian leaders frequently espouse anti-Western or anti-elite narratives.

Taken together, the demographic and regional analysis highlights that belief in the moon landing hoax is not just an isolated curiosity but part of a broader context of trust, education, political affiliation, and cultural narratives. This examination helps explain why the theory, despite being debunked numerous times, continues to resonate with a segment of the global population.

The Evolution of Public Opinion and the Role of Technology

Public opinion about the moon landing hoax has evolved dramatically over the past five decades, influenced in large part by technological advancements and the ways information is disseminated. From the post-Apollo era of whispered conspiracies to today's viral internet memes and YouTube exposés, each era has had its own mechanisms for sustaining the belief that NASA faked the moon landings. The evolution of this belief reflects broader trends in how society consumes, questions, and challenges information.

In the 1970s and 1980s, conspiracy theories regarding the moon landing were passed around like urban legends, shared in niche magazines, small gatherings, or fringe radio shows. Bill Kaysing's book, *We Never Went to the Moon: America's Thirty Billion Dollar Swindle*, reached a small but dedicated audience, bolstered by its blend of skepticism and sensationalism. The belief in the hoax gained a foothold among those who distrusted the government, particularly following high-profile events like the Watergate scandal and the Vietnam War, which eroded public faith in institutions. However, without the mass media's spotlight or widespread connectivity, these beliefs were largely isolated and grew slowly within small pockets of society.

The 1990s marked a significant shift as the early internet began to provide a platform for those skeptical of the moon landings. Websites dedicated to conspiracy theories sprang up, and forums allowed like-minded individuals to exchange ideas and "evidence" at a pace previously unimaginable. This period saw the moon landing hoax theory transform from an underground suspicion into a subject of wider public discussion. As more people went online, the theory attracted curiosity beyond its original base, reaching individuals who may not have previously encountered such alternative narratives. This was also when mainstream media started to take notice, occasionally covering these theories, which paradoxically lent them a veneer of legitimacy.

Television also played a key role during this era. In 2001, *Fox Television* aired a special titled *Conspiracy Theory: Did We Land on the Moon?*, which introduced millions of viewers to the hoax theory. Although the program was not a serious investigation and lacked credible evidence, it sparked public curiosity and debates that lingered long after the broadcast. The show planted seeds of doubt that continued to grow, illustrating the power of visual media to influence public perception. While NASA and scientific experts were quick to refute the program's claims, the impact was significant—this marked a turning point where the moon landing hoax transitioned from fringe to mainstream awareness.

The dawn of social media in the 2000s and 2010s accelerated the spread of conspiracy theories like never before. Platforms such as Facebook, YouTube, and Twitter provided fertile ground for misinformation to thrive. Videos with titles like "Moon Landing Hoax: 10 Reasons It Was Faked" gained millions of views and were shared at lightning speed. The reach of user-generated content, often unchallenged and presented with an air of authority, led to an environment where belief in the moon landing hoax could proliferate unchecked. YouTube, in particular, became a hub where amateur documentary-makers used selective footage, dramatic music, and pseudo-scientific commentary to convince viewers of the hoax's credibility. Algorithms further pushed these videos to users who had shown even a passing interest, creating echo chambers where skepticism became reinforced without exposure to debunking viewpoints.

Modern technology has also given rise to more sophisticated forms of misinformation. Deepfakes, doctored images, and highly edited videos blur the line between fact and fiction, making it more challenging for individuals to distinguish between genuine information and cleverly disguised falsehoods. This digital environment has bolstered a new wave of moon landing skeptics, many of whom cite older "evidence" alongside new interpretations crafted in the age of viral content.

Interestingly, alongside the growth of the moon landing hoax theory, a parallel trend has emerged: more extensive efforts to debunk the myths. NASA, prominent scientists, and educators have taken to these same platforms to explain how the Apollo missions were indeed real, using everything from archived footage to modern simulations of lunar conditions. Despite these efforts, public opinion remains split, with a persistent minority maintaining their belief in the hoax. The resilience of this belief underscores a broader societal shift toward questioning authority and prioritizing personal research over expert consensus.

In conclusion, the trajectory of public opinion about the moon landing hoax reflects an ongoing cultural and technological evolution. From whispered rumors to digital crusades, the moon landing conspiracy theory exemplifies how beliefs adapt and endure through changing media landscapes. Despite decades of scientific evidence, the theory persists, driven by a complex interplay of trust, access to information, and the human inclination to question reality itself.

CHAPTER 15

Chapter 15: The Scientific Method vs. Conspiracy T

The Foundations of the Scientific Method and Its Role in Truth-Seeking

To understand why the moon landing hoax theory continues to attract followers despite overwhelming evidence to the contrary, it is essential to contrast the methods used by scientists with those employed by conspiracy theorists. Central to this discussion is the scientific method—a systematic process that has been the cornerstone of scientific inquiry for centuries.

The scientific method begins with observation, where scientists note a phenomenon or raise a question. This is followed by the development of a hypothesis—a tentative explanation that can be tested. Rigorous experimentation and data collection come next, designed to challenge or support the hypothesis. The analysis phase is crucial, as it involves scrutinizing the data for patterns and inconsistencies. Finally, the conclusion is reached, and findings are presented for peer review, allowing other experts to replicate the study and validate its results.

This process fosters a culture of transparency and self-correction. Scientists build on each other's work, and their conclusions are only as strong as the evidence that supports them. The Apollo moon landings are a prime example of the scientific method in action. Every aspect of the missions, from the engineering of the Saturn V rocket to the biological resilience of astronauts, was meticulously planned, tested, and documented. These achievements were not isolated to NASA; they involved a global community of experts. Independent astronomers from multiple countries tracked the Apollo missions, while samples of moon rocks were distributed worldwide and analyzed by scientists who confirmed their extraterrestrial origin.

The moon landings were supported by an unprecedented amount of data, photographs, videos, and firsthand testimonies from astronauts and mission controllers. The evidence underwent scrutiny not just at the time of the missions but in the decades that followed. Peer-reviewed studies analyzed everything from the radiation exposure faced by the Apollo astronauts to the soil composition of the lunar samples. This rigorous scientific vetting established a solid consensus: the Apollo missions were real.

Moreover, the transparency of the Apollo program stands in stark contrast to the secrecy and speculation favored by conspiracy theories. Scientists thrive on challenges to their theories; every anomaly or question is an opportunity for deeper investigation. For instance, if a shadow in a photo appeared longer or oriented differently than expected, researchers would look for the reason, whether it be an uneven lunar surface or the effects of multiple light sources. This investigative rigor ensures that explanations are grounded in physical laws and empirical evidence.

In contrast, conspiracy theorists often bypass the process of peer review and controlled testing. Claims that the moon landings were staged typically hinge on surface-level observations—perceived anomalies in photos or supposed contradictions in testimony. These claims rarely stand up to detailed analysis. When experts dissect these allegations, they find that the so-called evidence is either misinterpreted or based on a misunderstanding of photographic techniques, physics, or space conditions. For example, the infamous "waving flag" argument overlooks the fundamental principle that the flag was designed with a horizontal rod to keep it unfurled, and the movement captured was the result of inertia from being planted into the lunar soil.

The scientific method does not promise infallibility, but it provides a robust mechanism for refining knowledge. Mistakes are corrected, and theories evolve based on new data. This adaptability is what solidifies the legitimacy of scientific conclusions, such as the successful Apollo missions. In contrast, conspiracy theories tend to resist adaptation; instead of adjusting to new facts, they expand their narratives to incorporate them as further proof of deception.

Understanding the role of the scientific method reveals why the claims of a hoax fall apart under scrutiny. It highlights the gulf between evidence-based reasoning and the selective narratives driven by conspiracy thinking, setting the stage for a deeper look into why such beliefs persist despite the clarity that rigorous science offers.

Characteristics of Conspiracy Thinking

While the scientific method is celebrated for its rigorous adherence to evidence and peer-reviewed validation, conspiracy thinking operates on a vastly different plane. To understand why moon landing hoax theories continue to persist, it's essential to explore the psychological and cognitive attributes that define conspiracy thinking.

At its core, conspiracy thinking relies on patterns of thought that prioritize intuition and suspicion over hard evidence. A hallmark trait is the belief that powerful, hidden forces are orchestrating events behind the scenes. For moon landing conspiracy theorists, this manifests in the idea that NASA, bolstered by the U.S. government, staged an elaborate hoax to deceive the world. This mindset is sustained by the assumption that apparent inconsistencies or anomalies must signal a cover-up rather than being explainable through simpler, evidence-based reasoning.

One of the primary features of conspiracy thinking is confirmation bias. This cognitive bias leads individuals to seek out and favor information that aligns with their pre-existing beliefs, while dismissing or ignoring evidence that contradicts them. For instance, a person convinced that the moon landing was faked might point to the peculiar appearance of shadows in the photographs as proof of studio lighting, while ignoring credible explanations related to the properties of light on the moon's uneven surface. The desire for evidence that fits neatly into a preordained conclusion drives the way information is gathered and interpreted.

Proportionality bias is another characteristic that fuels conspiracy theories. This is the belief that significant events must have equally significant causes. The moon landing was a monumental achievement—an event so awe-inspiring that some find it hard to believe it could have been executed without some grand deception. Conspiracy theorists argue that a mission as complex as Apollo 11 could not have succeeded with the technology of the 1960s, fueling doubts that lead to more elaborate theories of fraud.

Selective skepticism also plays a crucial role. In contrast to genuine skepticism, which applies critical thinking evenly across all claims, selective skepticism is unevenly distributed. Moon landing skeptics are highly critical of official narratives from NASA and the

U.S. government but will often accept dubious or unverified information that supports their theories with little scrutiny. For example, a claim suggesting that famed director Stanley Kubrick helped fake the moon footage might be embraced without evidence, simply because it bolsters the hoax argument. This selective approach undermines the balance required for credible analysis.

Moreover, conspiracy theories thrive on the notion that the "truth" is being deliberately obscured. This mindset often leads to the moving of goalposts—a strategy where, when one piece of the theory is debunked, proponents shift to another angle rather than revisiting their original conclusion. In the context of the moon landing, even after decades of evidence, ranging from lunar rock studies to satellite imaging of the Apollo landing sites, conspiracy believers may pivot to claims about NASA's manipulation of newer data or suggest that the evidence itself was faked later.

Conspiracy thinking also leans on anecdotal evidence and sensational claims rather than peer-reviewed studies or verified data. For example, proponents may argue that a whistleblower or a mysterious source has confirmed the hoax but provide no verifiable documentation. These anecdotal stories are often impossible to confirm and fall outside the scope of traditional scientific inquiry. Despite their unverifiability, they are compelling to believers who already view the official version of events with deep skepticism.

In contrast, scientists approach anomalies as puzzles to be solved, not proof of deception. If shadows appear out of place or an astronaut's movement seems counterintuitive, researchers employ knowledge of optics, physics, and space conditions to explain these phenomena logically. The comprehensive nature of scientific rebuttals often reveals that supposed anomalies are consistent with lunar conditions or photographic nuances.

The persistence of moon landing hoax theories demonstrates how conspiracy thinking operates outside the realms of evidence-based analysis. It draws from intuition, amplifies anomalies, and clings to alternative explanations even when confronted with overwhelming evidence. This dichotomy between the scientific approach and the structure of conspiracy theories illuminates why such beliefs, despite being debunked time and again, remain compelling to some. Understanding this contrast allows us to appreciate not just the moon landing debates but the broader human tendency to seek hidden explanations behind major historical events.

Bridging the Divide – Critical Thinking and Education

While the scientific method and conspiracy thinking occupy opposite ends of the evidence spectrum, understanding why people fall into conspiracy thinking and how to bridge the gap is essential for fostering informed discussions. The final key to this contrast is promoting critical thinking and educational strategies that empower individuals to assess information objectively.

Critical thinking is the process of evaluating information in an unbiased manner. It encourages individuals to analyze claims based on evidence, logic, and credibility rather than intuition or presupposed beliefs. By applying critical thinking, a person can move beyond surface-level observations and examine deeper connections that validate or refute a theory. For example, a critical thinker encountering the moon landing hoax theory would weigh the alleged inconsistencies in NASA's photos against the expertise of photographers and scientists who provide explanations rooted in known optical and physical laws.

One of the primary tools in promoting critical thinking is teaching individuals how to identify reliable sources. The internet has made information more accessible than ever, but it has also made it easier for misinformation to spread. Recognizing the difference between peer-reviewed studies and unsourced claims is fundamental. A scientific study explaining how astronauts navigated the Van Allen radiation belts, backed by detailed research and published in a respected journal, carries more weight than an anecdotal assertion on a conspiracy website. Education that highlights the value of verified sources over sensationalist claims can help individuals distinguish between substantive arguments and spurious theories.

Additionally, promoting a balanced form of skepticism—one that questions all claims, not just those from authority figures—strengthens the ability to navigate complex topics. True skepticism requires questioning any information, whether it supports a mainstream narrative or a counter-narrative. In the case of the moon landing, a balanced skeptic would analyze both NASA's official reports and the counterarguments from conspiracy theorists, seeking out data and logical consistency rather than leaning toward one side purely out of bias. This type of skepticism is what drives scientific progress: theories are tested, challenged, and improved upon rather than accepted uncritically.

Cognitive biases, such as confirmation bias and proportionality bias, can be mitigated by fostering self-awareness and mindfulness in information processing. When individuals understand their cognitive predispositions, they can actively counteract these biases by challenging their assumptions and seeking out conflicting information. For instance, if someone inclined to believe in the moon landing hoax encounters a compelling explanation that challenges their belief, a strong grounding in critical thinking will encourage them to consider it fairly rather than dismissing it out of hand.

Education plays a critical role in fostering these skills, especially in an era of increasing polarization and information overload. Integrating lessons on the scientific method, logical fallacies, and media literacy into school curricula can help students develop the tools they need to evaluate complex information. Teaching students to ask questions such as, "What evidence supports this claim?" or "What alternative explanations exist?" equips them to think independently and resist the allure of baseless conspiracy theories.

For adults, public outreach and science communication can bridge the knowledge gap. Scientists and educators can make their work more accessible by explaining concepts in straightforward language and addressing common misconceptions directly. For instance, explaining why the flag seemed to "wave" on the moon or why shadows behave differently on a non-atmospheric body can demystify arguments used by hoax proponents. Open, respectful conversations that address questions without derision encourage curiosity and engagement rather than alienation.

A strong society benefits from a population capable of discerning credible information from misleading claims. While conspiracy theories will always exist, promoting critical thinking and a thorough understanding of scientific inquiry provides a counterbalance that helps individuals resist the pull of unfounded beliefs. By bridging the divide between the rigorous standards of science and the often instinct-driven world of conspiracy thinking, we can cultivate a more informed and thoughtful public, capable of appreciating the wonders of achievements like the Apollo moon landings without succumbing to doubt rooted in misinterpretation or misinformation.

CHAPTER 16

Chapter 16: Final Thoughts – Why the Moon Landing

The Historical and Inspirational Significance of the Moon Landing

The Apollo 11 moon landing on July 20, 1969, was more than a singular feat of technological prowess; it was a defining moment that showcased humanity's ability to dream, innovate, and achieve the seemingly impossible. As millions of people across the globe watched Neil Armstrong take those first steps on the lunar surface, his words—"That's one small step for [a] man, one giant leap for mankind"—resonated as a powerful testament to the capabilities of human endeavor. This event marked not just a victory in the space race but a transcendent milestone that unified people in a shared sense of awe and possibility.

The moon landing symbolized the culmination of intense dedication, rigorous scientific pursuit, and an unyielding commitment to exploration. In the context of the Cold War, where geopolitical rivalries often manifested in military brinkmanship and ideological divides, the success of Apollo 11 stood as a beacon of peaceful achievement. It showed the world that competition could lead to remarkable contributions to human knowledge and that the quest for space was not just a contest but a pathway to shared advancement. This breakthrough invigorated a generation of scientists, engineers, and visionaries who saw in it a reminder that the boundaries of exploration were only limited by the collective will to transcend them.

Beyond its immediate impact, the moon landing continues to inspire. The idea that human beings could overcome daunting challenges to reach a celestial body remains an enduring narrative that fuels the imagination of future explorers. Young children who marveled at grainy images of astronauts bouncing on the lunar regolith often grew up to be the innovators, technologists, and space enthusiasts who propelled subsequent leaps in science. The moon landing's legacy is woven into the fabric of modern advancements—from space telescopes peering into the farthest reaches of the universe to missions aimed at colonizing Mars.

The inspirational ripple effect of Apollo 11 extends beyond the sciences. It informs art, literature, and philosophy by reminding people of a profound truth: humans are naturally driven to explore, to question, and to push boundaries. The sheer audacity of leaving Earth and landing on the moon underscores the power of human ingenuity and collaboration. It is a story of perseverance, where scientists and astronauts overcame obstacles ranging from technical failures to the unknowns of space. Their success was not guaranteed, but their triumph demonstrated what could be accomplished when intellect, skill, and determination converge.

Today, even as new space programs continue to evolve, the moon landing remains a touchstone of human achievement. It challenges contemporary society to think beyond immediate horizons, inspiring current space missions such as NASA's Artemis program, which aims to return astronauts to the moon and set the stage for deeper space exploration. The echoes of Apollo serve as a reminder that humanity's quest for knowledge is not just a series of historical footnotes but a continuous journey propelled by curiosity, resilience, and the drive to see beyond what is known.

In the end, the moon landing was not just a victory for the United States; it was a victory for humankind. It validated the belief that with enough dedication, even the most formidable challenges could be met and overcome. This spirit of exploration and achievement is why the story of the Apollo program continues to captivate, inspire, and affirm the human capacity for greatness.

The Influence of Conspiracy Theories on Public Trust

The moon landing hoax theory, while dismissed by the vast majority of scientists, historians, and rational thinkers, has nonetheless managed to carve out a niche within public discourse that undermines trust in institutions. The persistence of these conspiracy theories reveals a deeper tension between belief in empirical evidence and a skepticism that borders on mistrust of authority. The Apollo 11 mission, originally intended as a testament to human innovation and cooperation, is now also emblematic of how grand achievements can be clouded by doubt when misinformation spreads.

At the core of this skepticism lies a fundamental challenge: how to foster trust in science and government in an era defined by information overload and polarized beliefs. The moon landing conspiracy, despite being thoroughly debunked by photographic evidence, firsthand accounts, and technological analysis, persists because it taps into a deeper psychological phenomenon. People often question complex achievements when they feel disconnected from or distrustful of the organizations responsible for them. In the case of the Apollo program, the monumental scale of success—reaching the moon in less than a decade—seems implausible to those who do not fully grasp the technological progress of the era or the relentless work of the engineers and astronauts involved.

These theories are fueled further by the rise of the internet and digital media, which allow any narrative, regardless of its factual grounding, to find an audience. Online forums and social media have amplified moon landing conspiracies by presenting them alongside legitimate news and research, often blurring the line between fact and speculation. This environment can sow confusion, making it harder for the general public to discern which sources are reliable and which are not. When someone encounters multiple accounts that cast doubt on a widely accepted truth, even a baseless theory can gain traction as a plausible alternative.

The implications of this erosion of trust extend beyond the realm of space exploration. When people begin to doubt verified historical events, it sets a precedent for the rejection of other scientific and historical facts, breeding cynicism and mistrust in fields such as climate science, public health, and government initiatives. The moon landing conspiracy thus serves as a case study for how misinformation can persist and proliferate, influencing not just niche groups but mainstream perceptions. This has led to an environment where conspiracy theories about large-scale events thrive, often at the cost of faith in societal progress.

To understand why individuals might choose to believe in these theories, one must look at the psychological appeal of conspiracy thinking. Conspiracies often provide a sense of clarity in a chaotic world; they simplify complex events into narratives of betrayal and hidden truths. For those who feel disillusioned or powerless, accepting a theory that positions them as one of the enlightened few who "know the truth" can be empowering. This is why combating these beliefs requires more than just presenting evidence; it requires engaging with the social and emotional drivers that make conspiracy theories compelling.

NASA and the broader scientific community have worked tirelessly to counter these narratives with transparency and outreach. Educational initiatives, detailed documentaries, and public access to mission archives are just some of the ways these institutions aim to reinforce the veracity of the moon landings. Nevertheless, countering the belief in hoaxes requires an understanding that trust cannot be built solely through facts; it must be fostered through open dialogue, education, and consistent demonstration of integrity.

The Apollo program, once a symbol of boundless possibility, now also stands as a testament to the duality of human nature: capable of reaching the stars, yet prone to doubts that can pull us back to earth. Recognizing the dynamics behind these doubts can help address the wider challenge of how society handles misinformation, ensuring that truth prevails not only in the retelling of history but in its ongoing creation.

The Lasting Legacy of Apollo and the Balance Between Inspiration and Skepticism

Despite the shadow cast by conspiracy theories, the Apollo moon landing's legacy remains one of the most awe-inspiring achievements in human history. It embodies the triumph of collective vision, relentless pursuit, and the profound determination to transcend earthly limitations. Yet, as the debate over the authenticity of this milestone has demonstrated, the legacies of grand events often need safeguarding—not just against the passage of time, but against doubts that erode their significance. How society balances inspiration with healthy skepticism shapes how achievements are remembered and their impact on future generations.

The Apollo program did not merely pave the way for space exploration; it ignited a cultural and scientific renaissance that still echoes today. The moon landing showcased the boundless potential of human ingenuity, spurring technological advancements in everything from computing to telecommunications. The trajectory of innovation sparked by the Apollo missions has touched countless areas of modern life, underscoring why its legacy is critical to defend. When conspiracy theories take hold, they risk overshadowing these contributions and diminish the hard work and sacrifices of the people who made the impossible possible.

A central aspect of this legacy is the message that space is the next frontier, waiting not just for astronauts but for the dreamers, scientists, engineers, and artists who seek to explore its mysteries. The moon landing offered a glimpse into the boundless opportunities beyond our planet, inspiring new generations to continue humanity's quest for knowledge. Programs like NASA's Artemis aim to revive the spirit of Apollo, with plans to return humans to the moon and use it as a stepping stone for Mars exploration. This enduring pursuit of space exploration echoes the ethos of Apollo: pushing the boundaries of what we know to uncover new truths.

Yet, to maintain the inspiration sparked by these milestones, society must understand the underlying drivers of disbelief and distrust. Acknowledging the psychological and cultural reasons why some people subscribe to moon hoax theories is essential to fostering informed discourse. Efforts to address this challenge must go beyond the realm of rebuttal and delve into building public trust through open communication and genuine engagement. Strengthening scientific literacy and fostering critical thinking are essential tools to help people differentiate between healthy skepticism and baseless doubt.

In defending the authenticity of the moon landings, scientists, educators, and space advocates emphasize not just the technical evidence but the story of human collaboration and sheer determination that made the moon landings a reality. The achievements of the Apollo program belong to all of humanity, serving as a reminder that when we aim high and work collectively, we can overcome monumental challenges. Even in the face of skepticism, it is essential to preserve and celebrate this legacy, ensuring that the lessons and inspiration from Apollo continue to light the path forward.

As history has shown, moments of collective triumph are rare but powerful. They anchor us to the realization that humanity can reach beyond its grasp. The Apollo missions are one of those rare moments, a testament to what can be achieved when imagination and action align. Even as conspiracy theories continue to challenge the validity of such feats, the enduring lesson from Apollo is that the truth—rooted in evidence, courage, and shared human spirit—can withstand the test of time. The moon landing matters not only for what it achieved but for what it continues to symbolize: a vision of humanity that is curious, determined, and capable of extraordinary things.